GCSE PASSBOOK

BIOLOGY

June Hassall

First published 1988
by Charles Letts & Co Ltd
Revised 1990
Diary House, Borough Road, London SE1 1DW

Illustrations: Ann Savage and Tek Art

British Library Cataloguing in Publication Data
Hassall, June
Biology. —— (Key facts. GCSE passbooks).
1. Biology
I. Title II. Series
574 QH308.7

ISBN 0-85097-803-3

Printed and bound in Great Britain by
Charles Letts (Scotland) Ltd

Contents

Preface

This book is part of the Letts Keyfacts Passbooks series designed to provide a concise revision course for the GCSE. This book covers the basic topics of the Biology syllabuses prepared by the

London and East Anglia Group (LEAG);
Midland Examining Group (MEG);
Northern Examining Association (NEA);
Northern Ireland Examinations Council (NIEC);
Scottish Examinations Board (SEB);
Southern Examining Group (SEG), and the
Welsh Joint Education Committee (WJEC).

The material is arranged in the same order as the Themes identified for Biology in the National Criteria. The main topics are clearly identified and there is good use of labelled diagrams to aid in understanding.

The book not only deals with *what* to revise, but also *how* to revise it, with useful sections on how to get organized for revision and how to make a revision timetable. At the end of each section of work there are practice questions for students to try. These have answers or marking schemes which will show students what is required in the examination. There are also hints on how to do well on the practical skills assessment and the examinations.

It is hoped that the book will be useful not only to students in school but also to external or mature candidates. The examination boards make special provision for such candidates by providing written papers covering some of the practical skills. Reference should be made to the appropriate syllabuses.

It is a pleasure to thank Mr Paul Dickinson (Head of Biology at Sir John Gleed Boys High School) who has given much helpful advice and encouragement during the planning and writing of the book. I am very grateful for his contribution. I also thank Mr Dickinson's Biology students who have tried out the questions and given useful feedback. I would also like to express my appreciation to Mr Morton Jenkins for his careful reading of the manuscript and his constructive comments.

I am most grateful to the staff of Charles Letts and Co. Ltd. for their expertise and guidance during the preparation of this book.

June Hassall 1990

Introduction and guide to using this book

This book is going to help you to revise for your GCSE Biology examination. It has several parts which will show you what to revise and also *how* to revise it.

What you need to know and how you will be tested on it is covered on pages 6–7. Read this first so that you know what you will be doing.

Getting organized for revision is on pages 8–9. It tells you how to set up your Revision Book for Biology and how to let your classwork help you to plan for examination success.

Planning your revision on pages 10–11 shows how to set up a revision timetable and gives you more hints on how to revise.

The information and skills which you need to revise for your examination are on pages 13–140. They are divided into 25 chapters which are listed in the Contents. Twenty-four chapters cover the content of the syllabus, and the last chapter covers the practical skills. You should revise each chapter in turn. Important words and ideas are picked out in bold type or in capital letters to help you, and special triangular symbols in the margin point to key facts.

The chapters are grouped into sections and there are self-test questions at the end of each section. Do the questions after you have revised each section and then check your answers against the answers which start on page 141. When you add up your marks for each test you will be able to see how well you are doing. Do *not* peep at the answers until after you have done the questions!

The last part on final examination preparation on pages 157–8 gives some advice on last-minute revision and how to answer the various kinds of questions on your examination.

If you want to find information on a certain topic, or to do a last-minute check up, use the Contents page or use the alphabetical list of topics in the Index.

Now, the rest is up to you. Work hard and do well!

Introduction to GCSE Biology

The GCSE (General Certificate of Secondary Education) examinations in Biology have to follow certain guidelines (the National Criteria). These guidelines list the core content of the examination, which every syllabus must contain. This is listed below. The topics in this book are arranged in the same order as the four Themes of the National Criteria for Biology.

Themes of the National Criteria for Biology
Theme one: Diversity of organisms (5–10 per cent of the marks). This is the study of living things, mostly those in your own area. Simple keys are used to identify organisms. (In this book Theme one is covered in **Chapters 1 and 2** on pages 13–19.)

Theme two: Relationships between organisms and with the environment (25–40 per cent of the marks). This looks at how living organisms affect each other, and how they live in their environment. The effect that humans have on the environment is also included. (**Chapters 3–8** on pages 20–50.)

Theme three: Organization and maintenance of the individual (25–40 per cent of the marks). This begins with a study of cells, how substances move in and out of cells and how enzymes work. Then all the processes of living things are studied, in both plants and animals. (**Chapters 9–20** on pages 51–110.)

Theme four: Development of organisms and the continuity of life (15–25 per cent of the marks). This is about growth and reproduction in plants and animals. It includes the importance of chromosomes in passing characteristics from one generation to another and the ways in which characteristics can change. (**Chapters 21–24** on pages 111–134.)

How you will be tested

K **Knowing and understanding:** tested on the Theory papers, this covers the content of the four Themes. On pages 157-8 you will find hints on how to answer the various kinds of questions.

K **Practical and process skills:** tested on the Theory papers and by your teacher. These contribute 20 per cent to the final mark (30 per cent for NEA). The skills are listed on the next page, and more information is given on pages 135-140.

Practical skills needed

London and East Anglian Group (LEAG)

A Make and record accurate observations.
B Perform experiments and interpret results.
C Design and evaluate an experiment.

Midland Examining Group (MEG)

Following instructions
Handling apparatus/materials
Experimental design/problem solving
Recording/communicating
Observing/measuring
Interpreting data

Northern Examining Association (NEA)

1 Measurement
2 Observation
3 Handling apparatus/materials
4 Recording
5 Data and its interpretation
6 Design

Northern Ireland Schools Examination Council (NISEC)

I Manipulative skills
II Following Instructions
III Handling materials and measurement
IV Recording and presenting information
V Experimental design/problem solving

Scottish Examining Board (SEB)

Designing investigations
Recording results
Analysing results/drawing conclusions
Carrying out experimental procedures
Simple experimental techniques

Southern Examining Group (SEG)

1 To follow written and diagrammatic instructions
2 To handle apparatus and materials
3 To make and convey accurate observations
4 To record results in an orderly manner
5 To formulate an hypothesis
6 To design an experiment to test an hypothesis
7 To carry out safe working procedures

Welsh Joint Education Committee (WIJEC)

1 Observational and recording skills
2 Measurement skills
3 Procedural skills
4 Manipulative skills
5 Formulation of hypotheses and experimental design

Getting organized

Now, let's get down to work. We'll assume you've already read pages 5–7. You know how to use the different parts of this book, and you are probably a bit overcome by all the things that you'll have to learn! Well, where do you start?

1 Make a New Year's Resolution (even if it isn't January) to keep **your class notes tidy and to write up your practical activities properly**. You are going to need them both to study from. It is helpful if you underline important words in your notes and activities in colour so they will be easy to spot when you revise.

2 **Learn as you go along.** After each and every lesson read through your notes and make sure that you understand them. Read through the same topic in your textbook.

3 **Make yourself a revision book.** Buy an exercise book to use just for revising Biology.

4 . . . Some time later . . . All ready? Right, on pages 1 and 2 of your revision book copy out the names of the 25 chapters from this book, starting at

1. Different kinds of living organisms.

and going on to

25. Assessment of practical skills.

5 At the top of your page 3, write

Chapter 1 Different kinds of living organisms

Turn to page 13 in this book and read through that chapter. Now, **write into your own book the most important points**, e.g.

Characteristics of living things:
Respiration – release of energy.

(Your notes don't have to be proper sentences.)

6 When you have finished a chapter **read over your notes and then try to repeat them** from memory, a bit at a time. (Cover them over with a piece of paper as you go along.) Then look at

your class notes for that part and add in any other important points.
7 Now you've done Chapter 1, so put a tick by it at the front of your revision book.
8 Go on to the next chapter by writing its title in your book. Do the same things as before.
9 The next thing you'll come to are the Questions on Chapters 1 and 2. Turn to a clean page in your revision book and **write down the answers to the questions**.
10 Turn to the Answers section (which starts on page 141) and mark yourself. You may want to check your answers with your teacher or with a friend. **Learn from your mistakes.** Where you got something wrong, go back to the page where you wrote down the information and underline it with a coloured crayon or put a star * by it.
11 Now repeat these steps chapter by chapter. Where there are **diagrams and labels, copy them into your revision book**, study them and then try to do them again from memory.
12 Remember to tick off the chapters as you revise them.
13 When you have completed all the chapters you will be ready to **tackle some real examination questions!** Your teacher may have some of these or you may have to write to your Examination board or group (the Addresses are given on page 12) and purchase sample questions and past examination papers. Make sure that you only get the kind of papers for which you will be entered. There is no point in worrying over questions that you won't need to do.
14 Look at the instructions on the papers and work out the amount of time allowed for each question. In the examination you have to work to time, so it is good practise to do so now.
15 Find a quiet place where you will not be disturbed. Try to give yourself 'examination conditions' and do the questions in the same neat way that you would in an examination. There is no point in scribbling down notes: **write proper sentences** and draw labelled diagrams if these are needed.
16 Then, when you have finished the paper, you can use this book and your textbook to **check your answers**. If you are not sure if something is correct or not, then ask your teacher to help you with the marking.
17 You may find that certain parts of the syllabus, or certain kinds of questions give you trouble. Then you will know which areas need more work.

Planning your revision

'Revise' in the dictionary means 'to study again'. So you can't expect to learn something just by looking at it once. You have to write out the main points and go over and over them until you remember them. Making your own revision book is a good way to do this. Then you test yourself, and try again. Follow all the hints on pages 8-9. Revise all the time during your course, and then nearer the examinations make yourself a Revision Timetable.

Making your revision timetable

1 In your diary turn back 2 or 3 months before your first examination (depending on how much time you have left for revision). Let's say there are 80 days altogether. You may have other things planned, say for 6 days, leaving 74.
2 Now write the subjects you have to study into your diary, several for each day. Plan to study for about an hour, then have a 10-minute break, and so on. Don't do the same subject all day, but give yourself a change, for example

Friday	Daytime	School
	Evening	Biology Maths English
Saturday	Morning	English Biology Geography
	Afternoon	Sports
	Evening	Geography Disco

3 Now, for example in the parts marked 'Biology', write in the numbers of the Chapters from this book that you are going to revise, e.g. Chapters 1 and 2 on Friday evening, and Chapters 3–5 on Saturday morning.
4 Then, get started on your revision, and put ticks in your diary as you go through each part of the subject.

- Make your own summaries.
- Write out the main points on particular topics from memory.
- Draw and label important diagrams, and list the functions.
- Prepare tables comparing similarities and differences.
- Check over questions that you have already answered.

5 When you get your examination timetable write down in your diary the examinations you will be taking (put them in boxes).
6 Set aside the evening or morning before each examination for that subject. Write them in your diary. This is for your final revision, for example:

	Morning	Afternoon	Evening
Monday	Bio. revision	Bio. exam	Maths revision
Tuesday	Maths exam		

7 As the time for the Biology examination gets closer, go back over your own revision book and look at the questions you answered. Make sure you could now get them right.

Revision hints

Do keep your class notes tidy.
Do learn as you go along.
Do make your own revision book.
Do write down the main points.
Do make your revision timetable.
Do revise for several short periods.
Do test yourself by doing questions.
Do learn from your mistakes.

Don't leave your revision to the last moment.
Don't expect to learn something just by reading it once.
Don't get discouraged. Get help from a teacher or friend.
Don't keep on at the same subject for a long time.
Don't panic! There will be some questions you can do.

Address of Examination boards and groups

London and East Anglian Group (LEAG)
c/o University of London Schools Examination Board
Stewart House
London WC1B 5DN

Midland Examining Group (MEG)
c/o Oxford and Cambridge Schools Examination Board
Elsfield Way
Oxford OX2 8EP

Northern Examining Association (NEA)
c/o Joint Matriculation Board
Manchester M15 6EU

Northern Ireland Schools Examination Council (NISEC)
Beechill House
42 Beechill Road
Belfast BT8 4RS

Scottish Examinations Board (SEB)
Ironmills Road
Dalkeith
Midlothian EH22 1BR

Southern Examining Group (SEG)
c/o University of Oxford Delegacy of Local Examinations
Ewert Place
Summertown
Oxford OX2 7BZ

Welsh Joint Education Committee (WJEC)
245 Western Avenue
Cardiff CF5 2YX

1 Different kinds of living organisms

Aims of the chapter

After studying this chapter you should be able to:

1 Describe the characteristics of living organisms.
2 Identify the major groups of living organisms.

The characteristics of living things

All living things show certain characteristics which separate them from non-living things. These are listed below:

R for Respiration; release of energy from food.
E for Excretion: removal of waste products of metabolism.
M for Movement: within cells or parts of the organism.
I for Irritability: sensing changes and responding to them.
N for Nutrition: making or eating of food.
De for Development and growth: increase in size/complexity.
R for Reproduction: making of new organisms.

(NB. The first letters spell REMINDeR.)

Major groups of living organisms

Living organisms have differences from each other and these are used to classify them (divide them into different groups). There is a range of living organisms from the smallest and simplest to the largest and most complex.

BACTERIA e.g. Pneumonia. Cytoplasm. One chromosome but no nucleus. Cell wall. Very small. Saprophytes (decomposers) and parasites.

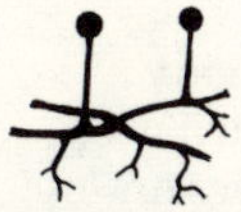

FUNGI e.g. Bread mould and Mushroom. Threads (hyphae). No chlorophyll. Form spores. Saprophytes (decomposers) and parasites. (**Lichens** are algae and fungi interwoven.)

VIRUSES e.g. Cold virus and AIDS. All are parasites and most cause disease. They are extremely small, have no cytoplasm or nucleus and only reproduce inside other organisms.
(NB. One-celled (unicellular) algae and protozoa can be grouped together as **PROTISTS**.)

PLANTS Chlorophyll. Cellulose cell walls. Stationary.

Algae e.g. Spirogyra. No stems, leaves or roots. One cell (unicellular) to many cells (multicellular). Simple threads or sheets of tissue. Mostly live in water.

Mosses. Simple stems, leaves and root-like structures. Reproduce by spores. Live in damp and shady places.

Ferns. Stems, roots and leaf-like fronds. Reproduce by spores. Live in moist and shady places.

Flowering plants Proper stems, leaves and roots. Flowers. Seeds inside fruits. Trees, shrubs and herbs.

Monocotyledons, e.g. Grass	*Dicotyledons, e.g. Daisy*
Long narrow leaves Veins in parallel lines One seed leaf (cotyledon)	Shorter, wider leaves Veins like a net Seeds with two cotyledons

ANIMALS No chlorophyll. No cellulose cell walls. Mobile.

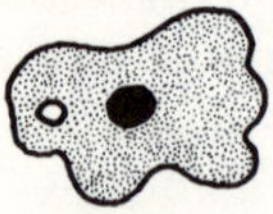

Protozoa, e.g. Amoeba. One-celled (unicellular). Irregular shape. Surrounds and feeds on smaller organisms. Reproduces by splitting into two halves.

Annelids, e.g. Earthworm. Ringed worms with many segments. Elongated. Moist skin. Bristles help in movement. Smooth band, the saddle (clitellum). Live in the soil.

Flatworms, e.g. Tapeworm. Flat body divided into many similar parts. Parasites. Make many eggs. Hooks and suckers on head.

Molluscs, e.g. Snails. Usually in a shell. Soft undivided body. Large muscular foot. Moist skin. Head with eyes and tentacles. Live in moist places.

Arthropods Segmented body. Jointed legs. External skeleton.

Arachnids e.g. Spiders. 4 pairs of legs. No wings. Head and thorax joined. Live on land.

Crustacea e.g. Crabs. 4–10 pairs of legs. No wings. Two pairs of antennae. Live in water.

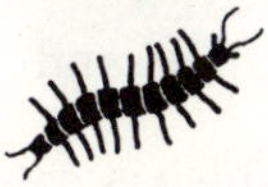

Myriapods e.g. Centipedes and millipedes. More than 10 pairs of legs. Body with many segments. Live on land.

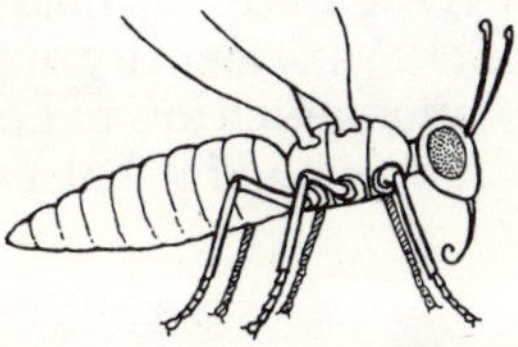

Insects e.g. Butterfly. 3 pairs of legs. Three body parts: head, thorax and abdomen. 2 pairs of wings. Antennae. Adults live on land, young forms may live in water.

Vertebrates Backbone of vertebrae. Internal skeleton.

Fish e.g. Goldfish. Fins. Gills. Scales. Gill cover (operculum). Live in water. External fertilization. Cold blooded.

Amphibia e.g. Frog. Moist skin. 4 limbs. Tadpole (water) and adult (land). External fertilization. Cold blooded.

Reptiles e.g. Lizard. Dry, scaly skin. 4 limbs. Live on land. Internal fertilization. Soft-shelled eggs. Cold blooded.

Birds e.g. Pigeon. Feathers. Legs with scales. 4 limbs. Wings. Beak. Internal fertilization. Hard-shelled eggs. Warm blooded.

Mammals e.g. Human. Hair. Mammary glands. 4 limbs. Internal fertilization. Young born alive. Warm blooded.

2 Using simple keys

Aims of the chapter

After studying this chapter you should be able to:

1. Understand how we name organisms.
2. Use a simple key to identify organisms.

How we name organisms

Organisms are divided into small groups called SPECIES. Members of the same species look similar. They can mate and produce offspring which are fertile.

Each small group has a 'double-barrelled' name which is written in *italics*. When you are using keys to identify organisms you will see these special scientific names which are in Latin so that everyone can understand them. So, instead of 'cat' or 'chat' or 'gato' everyone would use *Felis catus*.

Using a simple key to identify living organisms

Important external characteristics can be used to divide living organisms into smaller groups. For example the number of legs or the kind of body covering. Size is rarely used to separate groups, otherwise you might find you put the young and old organisms of the same kind into different species!

A specially prepared branching KEY can be used. At each point in the key you have to divide the organisms into two groups depending upon whether or not they have a certain characteristic. Let us look at an example:

In dividing the vertebrates into their main groups the first decision may be whether they have fins or not. This would separate off the fish.

1	(a)	With fins	Fish
	(b)	Without fins	Go to 2

Now the second characteristic may be whether or not they had wings. This would separate off the birds.

2	(a)	With wings	Birds
	(b)	Without wings	Go to 3

Then we would have with and without hair.

3	(a)	With hair	Mammals
	(b)	Without hair	Go to 4

And lastly; dry, scaly skin or smooth, damp skin.

4	(a)	With dry, scaly skin	Reptiles
	(b)	With smooth, damp skin	Amphibia

The key for vertebrates can also be laid out like this:

Vertebrates — with fins | Fish
without fins — with wings | Birds
without wings — with hair | Mammals
without hair — with scales | Reptiles
with smooth skin | Amphibia

When you use a key you are mostly being tested on your observation skills. You observe the external characteristics and use them to divide the organisms into smaller groups.

Summary: Different kinds of living organisms

1 Living organisms show the following characteristics: Respiration, Excretion, Movement, Irritability, Nutrition, Development and Reproduction. (REMINDeR is a memory aid.)

2 The main divisions of living things are Bacteria, Fungi, Viruses, Plants and Animals.

3 Plants are divided into non-flowering plants (Algae, Mosses and Ferns) and flowering plants (Monocotyledons and Dicotyledons).

4 Animals are divided into invertebrates (Protozoa, Annelids, Flatworms, Molluscs and Arthropods), and vertebrates (Fish, Amphibia, Reptiles, Birds and Mammals).

5 Living things can be classified into groups by using keys.

Test yourself

When you have studied Chapters 1 and 2 you will be ready for the first set of questions.

You should write the answers out in your Revision book. Make sure that you write down the question number and the part of the question that you are answering. For this first set of questions, tables have been made for you to copy. Copy them into your book and then write the answers into them. For later questions you will have to make your own tables in a similar kind of way.

When you are ready turn over the page and start!

Questions on Chapters 1 – 2

1 List the seven characteristics of living organisms. Write ONE sentence about each one to show that you understand what it means.

Questions **2–5** have five possible answers, (A),(B),(C),(D) or (E). For each question choose the best response. A response may be used once, more than once or not at all.

(A) Fish (B) Amphibia (C) Reptiles (D) Birds (E) Mammals

Which of the vertebrate groups listed above:

2 have fins? ____

3 have a body covering of scales and four limbs? ____

4 suckle their young? ____

5 have a tadpole stage in their life cycle? ____

6 The figure below shows four invertebrates.

(R) (S) (T) (U)

(a) Complete a table like that shown below to list 4 differences that can be seen between specimens (S) and (T).

		(S)	(T)
Differences	(i)		
	(ii)		
	(iii)		
	(iv)		

(b) Complete a table like that shown below to list 3 differences that can be seen between specimens (R) and (U).

		(R)	(U)
Differences	(i)		
	(ii)		
	(iii)		

(c) (i) Which 3 organisms of (R), (S), (T) and (U) do you think are most closely related? ____ ____ ____

(ii) List two characteristics which these organisms have in common which are NOT shared by the fourth organism.

I ______________________________

II ______________________________

7 Use the key below to identify the following six leaves.

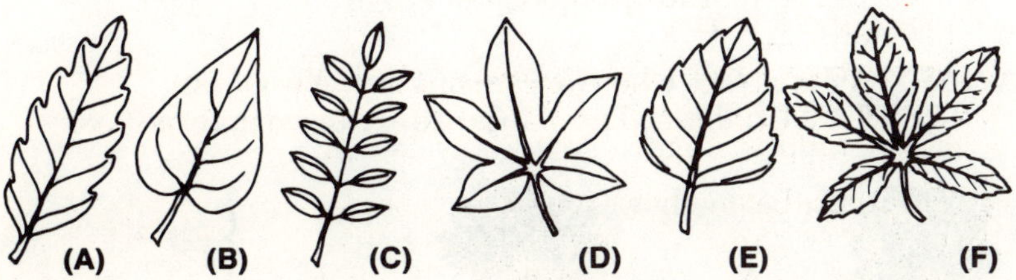

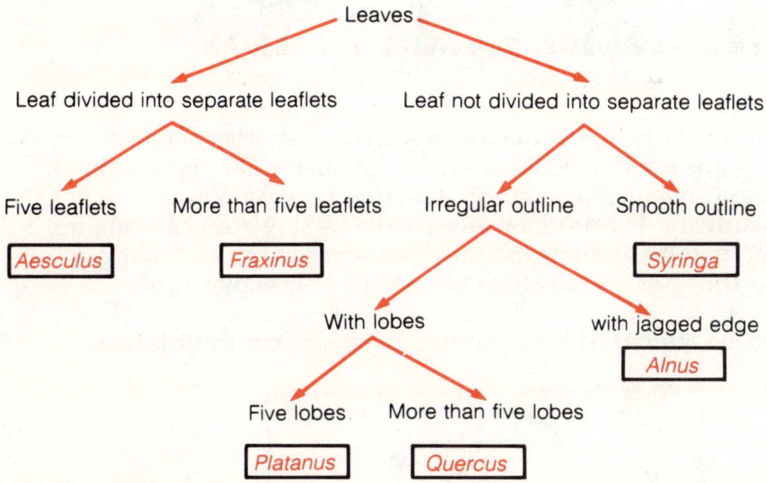

Record the names of the leaves

(A) (B) (C)

(D) (E) (F)

Now turn to page 141 to check your answers 1–7.

3 Organisms and where they live

Aims of this chapter

After studying this chapter you should be able to:

K
1 Describe the parts of a terrestrial and aquatic habitat.
2 Define population, community, environment and ecosystem.
3 Describe how organisms are collected and sampled.
4 Describe the effect of the environment on organisms.

Describing some habitats

K

A HABITAT is a place where organisms live. A wood is a TERRESTRIAL habitat. The plants are arranged in three layers.

There may be a **population** of oak trees, another of hazel shrubs and another of herbs such as bluebells. The plant, animal, bacteria and fungi populations together make up the **community**. The **environment** is the climate, the soil, and the water (non-living part) and the other organisms (living part). Everything interacting together is called the **ecosystem**.

A pond is an AQUATIC habitat and it also has three layers.

There may be **populations** of duckweed floating on the surface, tadpoles swimming in the water and pondweed rooted at the bottom. The plant, animal, bacteria and fungi populations make up the **community**. The **environment** is the water, soil and other organisms. Everything together is the **ecosystem**.

Sampling organisms

We may want to know how many organisms of a certain species there are in a habitat. It would take too long and be too difficult to count each one. So we count the number in a small part that has been chosen by chance – this is called a SAMPLE. We count the number of organisms in several samples and then find the average. From this we can estimate how many there would be in the whole habitat.

Methods used to collect and sample animals

Pitfall trap: for insects and other small organisms

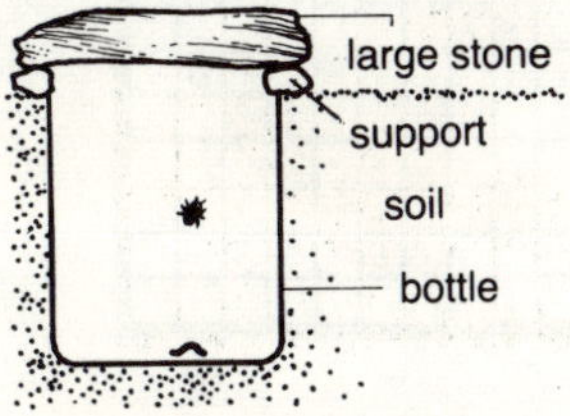

A bottle or tin is completely buried in the soil in the area to be sampled. Two small stones are used to support a large stone which is put on top to stop the rain getting in. Small animals walking in the area fall into the jar and are trapped.

Tullgren funnel: for small worms, e.g. nematodes and other small animals.

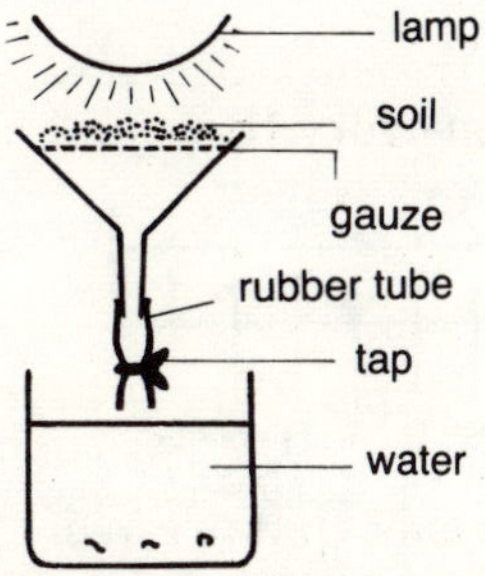

Some soil and the decaying remains of leaves (leaf litter) is collected from the area to be sampled. This is put on a wire gauze under a lamp. Any nematode worms move away from the light, and from the drying effect of the lamp. When the tap is opened they fall into the water and can be counted.

Nets: for aquatic organisms, and on land for insects.

Aquatic: A net is used to catch large organisms from the water. The same number of 'net-fulls' are used in different habitats. (A certain volume of water can also be examined.)

Land: A net (sweep net) can be pulled through long grass to collect insects such as grasshoppers. Another one (butterfly net) can be used to catch insects from the air.

Methods mainly used to sample plants

Quadrats: used in grassland, on rocks or on a tree trunk.

K

A square frame of wood or metal (QUADRAT) is placed anywhere in the habitat (i.e. at random). The NUMBER of plants of each kind in the quadrat is counted, e.g. certain species of weed. (Animals can also be counted, but they tend to move from place to place.)

A quadrat can be sub-divided by string or wire into 100 small squares. This is used when it is difficult to count individual organisms. Instead, we find the AREA occupied. Plant A occupies 9 parts out of 100, or 9 per cent. This is its PERCENTAGE COVER. Organism B occupies 15 parts, plus $4 \times \frac{1}{2}$ $= 15 + 2 = 17$ per cent of the area.

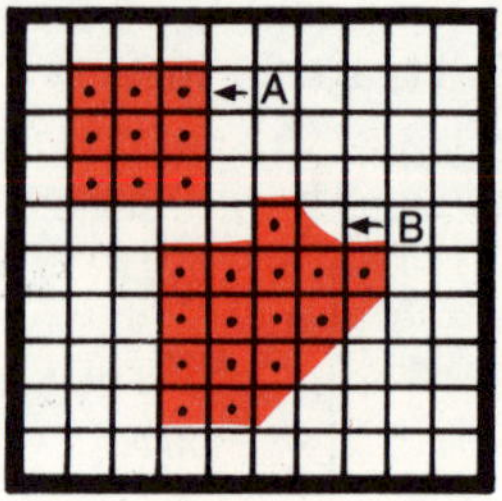

The percentage cover can be found for example for patches of grass on a path, of barnacles on a rock on the seashore or of algae such as *Pleurococcus* or lichen on a tree trunk.

Transects: used in grassland, seashore or ponds.

Line transect: A string is marked, for example, at every 10 cm and is stretched across the habitat. The plants found along the LINE near to each mark are recorded. (Any animals can also be recorded.)

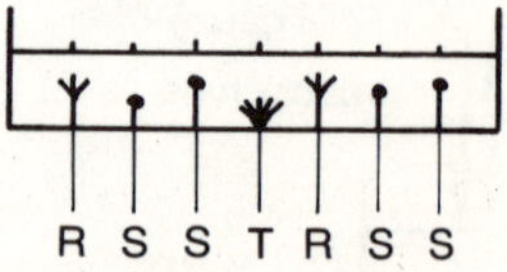

K

Belt transect: This is when you record the organisms that are in a wide BELT or band across the habitat. Quadrats can be placed along a line transect, and the organisms recorded.

K

POPULATION DENSITY is the number of organisms per unit area, e.g. number in a square metre on land, or per unit volume, e.g. number in a cubic metre in water.

PERCENTAGE COVER is the percentage area (parts out of 100) occupied by an organism.

Collecting organisms

You may be asked to describe how organisms are collected. These are some points you should remember:

1 Take only what you need.
2 Do not disturb the natural habitat.
3 Take a sample of the food eaten by the animals.
4 Put organisms in labelled containers to take to the lab.
5 Return animals unharmed to the habitat if possible.

In a terrestrial habitat:

1 Cut small pieces from flowering plants (don't pull up the whole plant!).
2 Use nets to sweep in long grass to find insects.
3 Use a knife to scrape snails or lichens from rocks.
4 Collect some soil and decaying leaves (leaf litter) to search for nematode worms, earthworms and millipedes.
5 Collect pieces of rotten wood which may have fungi growing on them.

In an aquatic habitat:

1 Use a net to collect surface organisms such as duckweed and water boatmen, and fish from the water.
2 Cut small pieces from flowering plants and pondweed.
3 Collect some of the bottom layer (detritus) and also any rocks or stones which have living things growing on them.
4 Collect some of the water to examine under a microscope for very small (micro-) organisms.

Summary: Parts of the ecosystem

1 A group of organisms of the same species in a certain place is called a population.
2 A group of several populations which interact with each other in a certain place is called a community.
3 The place where organisms live is called their habitat.
4 A habitat on land is a terrestrial habitat.
5 A habitat in water is an aquatic habitat.
6 The environment is the conditions in a certain place. These conditions are non-living (the climate, water and soil) and living (other organisms).
7 The ecosystem is the organisms and the environment as they interact with each other.

The effect of the environment on organisms

The ENVIRONMENT or ENVIRONMENTAL FACTORS are the conditions in a habitat. They are either:

K

(a) ABIOTIC FACTORS (non-living) such as climate (rainfall, light, temperature) and soil conditions (mineral salts, pH, soil moisture), or

(b) BIOTIC FACTORS (living) such as what an organism feeds on, what it eats and what decays its dead remains (see page 26).

K

The effect of light on plants

Light is important for plants because they need it to carry out PHOTOSYNTHESIS (making of food). Because of this:

1 Plants will bend or grow towards the light coming from one direction (PHOTOTROPISM, see page 95).
2 Moving unicellular algae (or plant-like protists) will swim towards the light.
3 Trees in a forest will grow as tall as they can to get a good supply of light.
4 Plants in the lower layers in a forest may be climbers on tree trunks to get further into the light.
5 Plants in the shade may grow very tall and yellow (become ETIOLATED) in their search for light.

Notes: Plants in the lowest layer in a forest and in a pond receive less light and have less photosynthesis and growth.

Some plants, e.g. mosses and ferns, prefer shaded conditions because these are cooler and more moist. So there is less chance of them drying out.

The number of hours of daylight controls the time at which some flowering plants start to flower.

The effect of light on animals

K

Some animals, e.g. woodlice, earthworms, nematode worms and cockroaches, move away from the light. This is because light conditions are usually also warm and cause loss of moisture which could make them dry out and die.

The effect of moisture on animals

Thin-skinned invertebrates, e.g. woodlice and insect larvae, will choose damp conditions in their natural habitats, and in the laboratory if put into a CHOICE CHAMBER (see page 96).

Measuring abiotic factors

Temperature. Use a −10°C to 110°C **thermometer**. Leave it in the habitat for at least 3 minutes before reading the scale so it has had time to come to the same temperature as the surroundings. The temperature in an open sunny area will be higher than in a shaded area. If the wind is blowing strongly it is likely to be cooler. It will also be cooler early in the day. There are also changes in temperature with the seasons.

Light. A **light meter**, like that used by photographers, is used to measure light intensity (the brightness of the light). Light intensity is highest in open sunny areas, and lowest in shaded conditions and when there are a lot of clouds.

The number of hours of sunlight is also important. There is less daylight in the winter than in the summer.

Water. We can measure the amount of rainfall using a **rain gauge**. This is buried in the soil. The funnel collects rain water from a certain area and this runs into the measuring cylinder. The amount of water in the measuring cylinder is read at intervals. A calculation is done to change it to a rainfall reading. The rain gauge is emptied and used again.

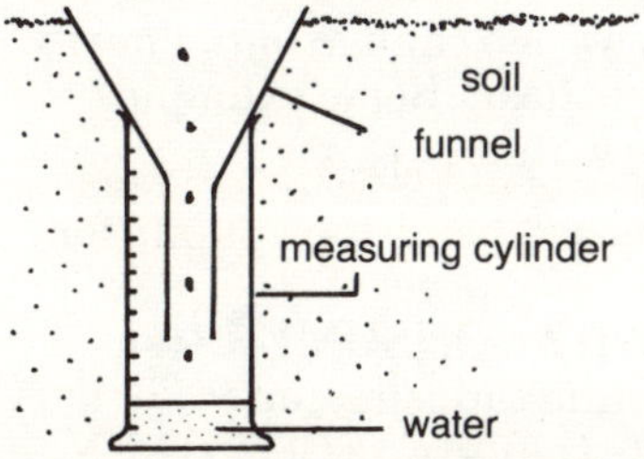

(See also page 36 for environmental factors in the soil.)

Summary

1 The ecosystem consists of the community of organisms interacting with the environment in the habitat.
2 Within a habitat several layers of organisms can be identified. These have different adaptations.
3 Animals can be collected and sampled using pitfall traps, Tullgren funnels and nets.
4 Sampling methods mostly used for plants are quadrats, line transects and belt transects.
5 Organisms are described by their density (number per square metre) and percentage cover (percentage area).
6 Organisms respond to abiotic factors in the environment such as light and water.

4 Food chains and food webs

Aims of the chapter

After studying this chapter your should be able to:

1 Define and give examples of producers, consumers, decomposers, food chains and food webs.
2 Draw food chains to show feeding relationships.
K **3** Interpret a food web and predict the effects of the death of some of the organisms.
4 Describe and explain the pyramids of numbers and biomass.
5 Explain the advantages of shortened food chains.
6 Describe some examples of biotechnology.

The feeding relationships of organisms

Organisms in a habitat are interrelated because of how they feed, and how their dead remains are decayed. Organisms can be divided up into:

K PRODUCERS. These are green plants which change light energy to trapped chemical energy by photosynthesis. Some examples are grass on land, and duckweed in the water.

K CONSUMERS. These are animals which depend upon plants for food. If they:
1 Eat plants directly they are PRIMARY CONSUMERS (or herbivores, such as grasshoppers or young tadpoles), or
2 Eat animals which have eaten plants they are SECONDARY CONSUMERS (such as small birds or small fish), or
3 Eat animals which have eaten animals they are TERTIARY CONSUMERS (e.g. large birds or large fish).

Secondary and tertiary consumers are also called carnivores. The highest level carnivore, e.g. hawk, is the top carnivore.

K DECOMPOSERS. These are organisms which decay dead organisms to release their nutrients. They are also called SAPROPHYTES (as they digest their food externally). Examples are fungi and bacteria.

FOOD CHAINS show which organisms feed on which. All food chains begin with a producer.

e.g. grass $\xrightarrow{\text{eaten by}}$ grasshopper $\xrightarrow{\text{eaten by}}$ small bird

FOOD WEBS show the interrelated food chains in a habitat. They can be very complex.

Some examples of food chains

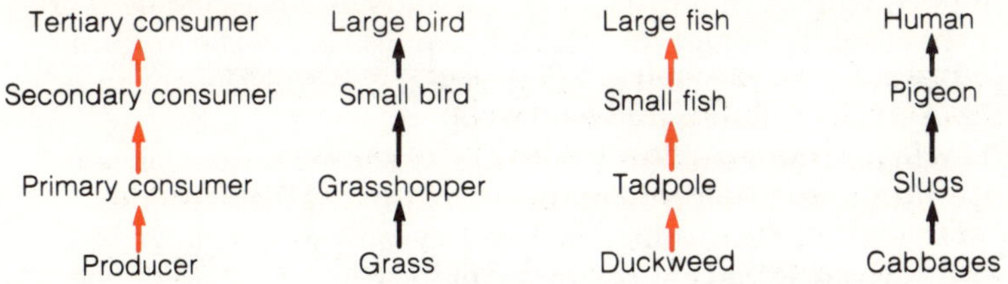

Note that the grasshopper is the PREY (it is eaten) and the small bird is the PREDATOR (it eats the other animal). But in turn the small bird becomes the prey (it is eaten by the large bird which is its predator).

The dead remains of all the organisms are decayed by bacteria and fungi.

A simple food web

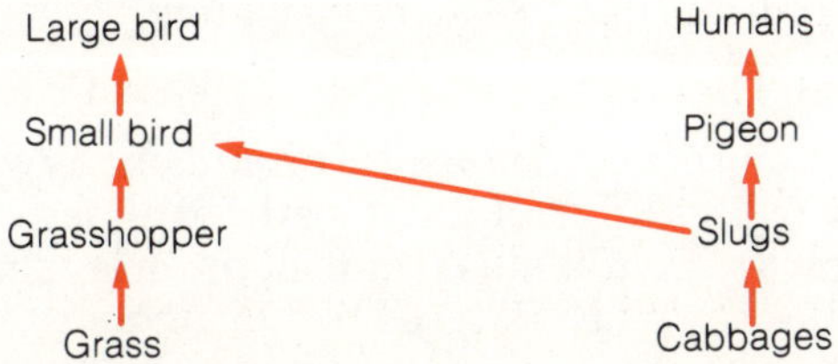

Try to imagine what would happen . . .

(a) If all the slugs died. Well, the cabbages would grow better as they would not be eaten. But the pigeons would have lost their food and they might start to die unless they could find something else to eat, . . . or

(b) If all the grasshoppers died. The grass should grow better because it was not being eaten. But the small birds would have to eat more slugs. And in doing this they would be in competition with the pigeons for food, . . . or

(c) If all the pigeons died. Well, humans could eat lots of other kinds of food. But the slugs might increase in numbers unless the extras were eaten by the small birds.

Try to make some of your own predictions from the food web.

Energy flow through living organisms

Primary consumers eat producers. But only about 10 per cent of the trapped energy from the plants becomes part of the animal's body. The other 90 per cent is lost in respiration and loss of body heat and all the activities of 'staying alive'.

When primary consumers are eaten by secondary consumers, again only about 10 per cent of the energy in the food becomes part of the organisms at the next level. So now we are down to 1 per cent of the original energy in the plants.

K At the next feeding level (or TROPHIC level, as it is called) this is reduced further to 1/1000th or 0.1 per cent of the original energy in the producers.

So there is less and less energy passed on as we go along a food chain. This leads to two things:

K **1** Less numbers of organisms can be supported as we go along a food chain. If we relied only on cabbages, there would be most cabbages, fewer slugs, fewer still pigeons, and fewest of all humans.

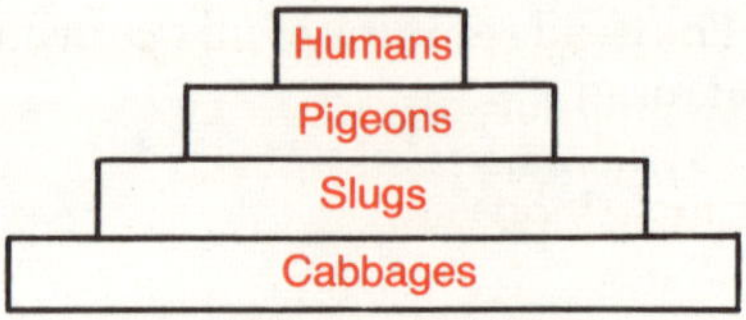

This is called the PYRAMID OF NUMBERS

K **2** Less total body mass (biomass) of organisms as we go along a food chain. So in a certain habitat if we dried all the cabbages they would have a greater mass than all the dried slugs, and so on. This is called the PYRAMID OF BIOMASS.

Shortened food chains

There is more trapped energy (food) and more total mass lower down in a food chain. And so:

1 We would get more energy by eating plants, than by eating animals that have eaten plants. This is one advantage of being a VEGETARIAN or VEGAN. We must be sure, however, that we get all the nutrients needed for healthy growth.

2 We could eat animals lower down in the food chain, herbivores (primary consumers) instead of carnivores (secondary consumers). We may not fancy the thought of eating slugs – but humans do eat snails, rabbits, chickens and small fish which are primary consumers. (A cow is also a primary consumer but it is very inefficient at making protein (beef and milk) from grass. Pigs and chickens are about three times better than cows at making protein from carbohydrates.)

Biotechnology

BIOTECHNOLOGY is the application of biology to industry for the making of foods, hormones and drugs and for getting fuel from wastes. The fact that living organisms can carry out certain processes is made use of, and multiplied to make it more efficient. The actual nature of organisms can also be changed by changing their chromosomes. This is called GENETIC ENGINEERING (see page 129) and it is part of biotechnology.

Foods. Micro-organisms reproduce very rapidly and they are kept in ideal conditions in FERMENTERS (where they grow with little or no oxygen). They grow and produce proteins:

1 **Single-celled protein** (S.C.P.) Bacteria are grown on energy-containing methanol made from North Sea gas. They are also given ammonia and mineral salts. They grow and reproduce rapidly and some are removed from the fermenter and dried and squashed to make foods like Pruteen which is protein rich and can be used to feed animals.

2 **Myco-protein.** This is made by a microscopic fungus in a fermenter. It is fed on a sugar solution made from wheat or maize, and also given ammonia to supply the nitrogen needed to make proteins. At intervals, fungus is filtered off and can be used as food. It is a more energy efficient way of making protein than getting it from cows, pigs or chickens.

3 Beers, wines, bread, yoghurt and cheese can be made by fermentation by yeasts (see also page 84).

Hormones. Insulin, needed by diabetics, can now be made by bacteria (see page 129).

Drugs. The antibiotic penicillin is made by growing the fungus *Penicillium* in a food source in large fermenters.

Summary

1 All food chains depend upon producers which trap energy.
2 Food webs are interwoven food chains consisting of producers, various levels of consumers, and decomposers.
3 Energy is lost at each stage of a food chain.
4 Fewer numbers, and less total mass can be supported the higher we go up a food chain.
5 Shortened food chains make better use of energy.
6 Humans use bacteria to produce food, hormones, drugs and fuels. This is part of biotechnology.

5 Recycling of materials

Aims of the chapter

After studying this chapter you should be able to:

1 Identify carbon, nitrogen and water as the basic raw materials needed by living organisms.
2 Distinguish between the re-use (recycling) of these raw materials and the one-way flow of energy in living things.
K **3** Describe the role of decomposers in recycling raw materials to be re-used by producers.
4 Describe the processes involved and the importance of the carbon, nitrogen and water cycles.
5 List ways in which water is essential for living things.

Raw materials and how they are recycled

Carbon, in the form of carbon dioxide, and water are combined by plants with energy supplied by sunlight to form carbohydrates and fats. With the addition of nitrogen (usually supplied as nitrates) plants can make proteins. These foods made by producers supply all the other organisms in the food chains. Carbon, water and nitrogen are therefore the basic raw materials of life.

'Food' can be thought of as consisting of two parts:

1 The simple chemical substances (raw materials): carbon, nitrogen and water.
2 The energy which holds these together to make energy-rich compounds.

K In the food chains the energy is gradually lost as we go from one feeding (trophic) level to the next. There is a ONE-WAY FLOW OF ENERGY. The energy in tertiary consumers cannot be recycled for producers to use. For living things to continue living there must be a continual addition of energy in the form of sunlight which is trapped by producers during photosynthesis to make more energy-rich compounds. Animals rely directly or indirectly for their energy on plants, and plants get this energy from sunlight.

K The raw materials (carbon, water and nitrogen) in the food also pass from one level to another in the food chain. But when all the organisms die, or when they release waste materials, these are decomposed and the raw materials are released and can be used again. MATERIALS ARE RECYCLED. Decomposers (certain fungi and bacteria) in the soil recycle the materials.

The role of decomposers

Saprophytic fungi such as moulds produce digestive juices onto the food source which makes it decay and become soluble (EXTERNAL DIGESTION). In a similar way saprophytic bacteria decay dead organisms and waste materials. In this way raw materials are recycled.

The carbon cycle

Plants contain carbon compounds, for example:

1 Green plants on land; in leaves, nectar used by butterflies, and storage organs used as food by animals.
2 Plants such as pondweeds, seaweeds, and plankton in the water.
3 Decaying plant remains like fallen leaves and dead plants in the leaf litter on the soil.
4 Dead plant remains of long ago which have become fossil fuels, e.g. coal and peat.

All these carbon compounds, and those from animals, can be COMBINED WITH OXYGEN to produce carbon dioxide and energy, either by respiration in living organisms, by decay or by the burning (combustion) of fuels.

But all these processes use up oxygen and produce carbon dioxide. There is a reverse process, PHOTOSYNTHESIS, which uses up carbon dioxide and releases oxygen. So the carbon dioxide and oxygen are recycled.

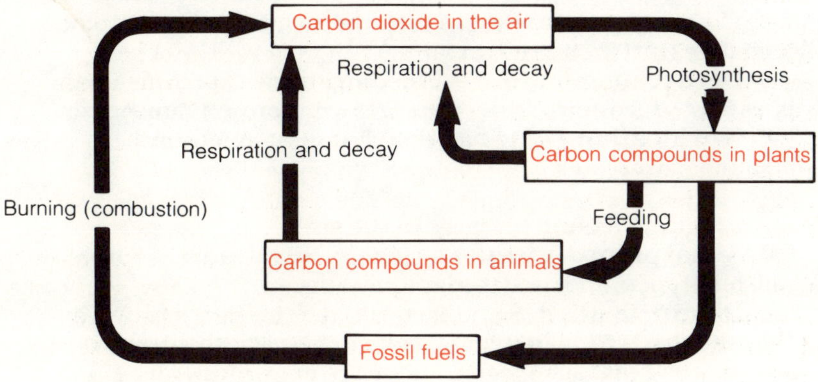

The nitrogen cycle

K

Nitrogen is needed to make proteins which are necessary for living things. Plants take in nitrates from the soil and make proteins. These are eaten by animals. Dead plants and animals are decayed by fungi and bacteria to release nitrogen compounds into the soil which are then changed to nitrates.

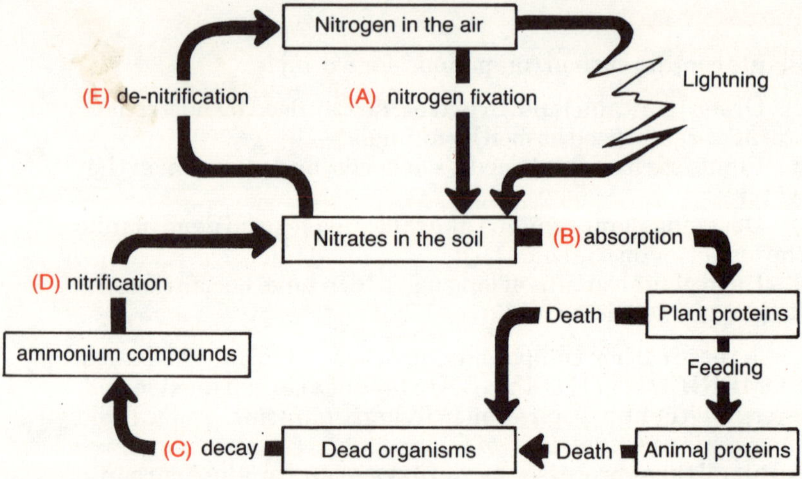

Processes increasing nitrates in the soil:

(A) **Nitrogen fixation.** Bacteria live in the bumps (nodules) on the roots of beans and peas (leguminous plants which produce pods). They change nitrogen gas in the air to nitrates. (Lightning also changes nitrogen to nitrates.)

(C) **Decay.** Fungi and bacteria decompose dead organisms and wastes to produce ammonium (nitrogen-containing) compounds.

(D) **Nitrification.** Other bacteria change ammonium compounds to nitrates.

Processes decreasing nitrates in the soil:

(B) **Absorption of nitrates** by plants. Nitrates are absorbed by the plant roots and used to make proteins.

(E) **De-nitrification.** Some bacteria work in the opposite way to the other bacteria. They change nitrates back into nitrogen. They can work without oxygen, and so are more numerous in soil with a lot of water in it (waterlogged soil).

Humans and the nitrogen cycle

In a natural habitat, such as a forest, the leaves fall to make leaf litter which may be taken into the soil by earthworms. Leaf litter, animal wastes and dead organisms are decayed by saprophytic fungi and bacteria, and the nitrates are recycled.

But when humans clear away the forest and use the land to grow crops they remove some of the nitrates and so reduce the FERTILITY of the soil (its ability to grow healthy plants). They would need to add extra nitrates to overcome this problem – but not so much that they make new problems!

Humans remove nitrates:

1 When the trees and shrubs are removed to make farmland, the plant proteins inside them are also removed.
2 When the ground is burned to clear it, the small plants and leaf litter are destroyed which might have decayed to return their nitrates to the soil.
3 When crops are grown, they take nitrates from the soil, and when humans reap and remove the crops they are taking away all the nitrates and other minerals inside them.

Humans add nitrates:

1 When they intentionally grow leguminous crops such as beans, peas and clover, and they leave the roots to decay in the soil. The nodules in the roots contain nitrogen-fixing bacteria which change nitrogen into nitrates.
2 When dried grass and other plants ('mulch') are added to the soil surface. The mulch protects the soil from drying out, and it decays to add nitrates to the soil.
3 When manure (excreta from farmyard animals), peat and compost (partly decayed plant remains) and humus (fully decayed plant remains) are dug into the soil. These are ORGANIC
K FERTILIZERS and they release nitrates slowly into the soil.
4 When INORGANIC FERTILIZERS (chemicals containing nitrates and other mineral salts) are added to the soil. The minerals are quickly released into the soil.

If more nitrates are added to the soil than the plants can use, some may be washed off the land into freshwater where they may cause a pollution problem (see page 39). So when humans interrupt the natural cycle they have to take care that they understand what this may lead to.

The water cycle

Water is necessary for life. It is recycled through living things, and after it rains it evaporates into the air.

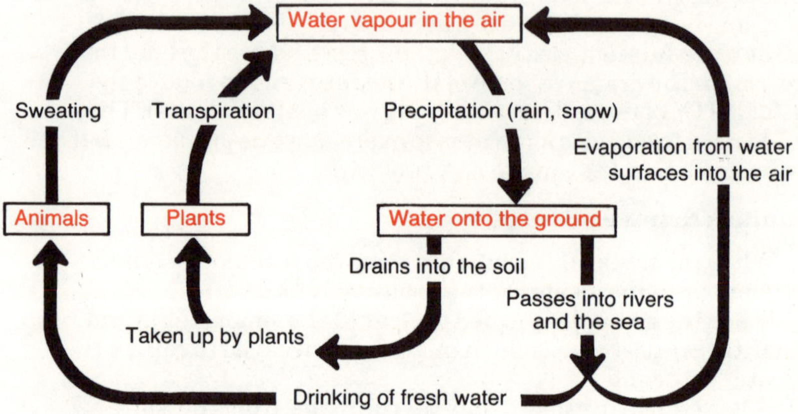

Using the diagram above describe how (i) plants, and (ii) animals help to recycle water.

The importance of water

1 Part of **protoplasm** (about 70 per cent) which is the living material of plant and animal cells.
2 As a **solvent** (liquid in which other substances dissolve): for example gases and digested foods dissolve in water. Many chemical reactions occur in water.
3 Takes part **in chemical reactions**, such as breakdown of foods during digestion, and of plant food stores during germination. A raw material in photosynthesis.
4 For **transport:** as the liquid part of the blood, and for transport in the xylem and phloem in flowering plants.

Summary

1 There is a one-way flow of energy through food chains. More energy from the sun must continually be added.
2 Raw materials (carbon, water and nitrogen) can be re-used (recycled) by the action of decomposers and other organisms.
3 Humans affect the carbon, water and nitrogen cycles.

6 Humans and the environment

Aims of the chapter

After studying this chapter you should be able to:

1 Describe examples of land use and possible problems.
2 Describe soil components and soil types.
3 Give ways in which soil fertility can be improved.
4 Describe the causes, effects and control of examples of air, land and water pollution.
5 Describe the need for, and examples of conservation.
6 Describe the advantages of recycling sewage.

Land use

Some ways in which humans damage natural habitats are:

1 DEFORESTATION: the clearing of land to be used for agriculture, or building houses and roads. (See also page 33.)

(a) Removing plants also affects the water cycle (see page 34), as the roots no longer take up water and return it to the air. This in turn can reduce the amount of rainfall.
(b) Burning wood and coal reduces the amount of oxygen and increases the amount of carbon dioxide in the air. The carbon dioxide layer can trap heat energy ('greenhouse effect').
(c) Trees take a long time to grow back again.
(d) Uncovered soil may be washed away.

2 Causing SOIL EROSION: washing away of fertile topsoil with decaying leaf litter, when it is no longer held together by plant roots and protected from heavy rainfall by the canopy of leaves.

3 MONOCULTURE: growing single crop species over wide areas of land, and from year to year. This uses up the soil nutrients as the crop keeps taking the same minerals. Pest species can also increase to high numbers as they can easily move from plant to plant of the same kind.

4 Using PESTICIDES: chemicals which kill insects (insecticides) and weeds (herbicides). Pesticides can cause pollution, and killing pests can upset the balance of nature. For example, if rabbits are killed because they are pests, then the foxes which used to eat them may now attack the farmer's chickens.

5 By adding pollutants (see pages 38–9).

Components of the soil

K

1 Humus: decayed remains of plants and animals, which helps soil to clump together into soil crumbs and contains mineral nutrients for plant growth.

K

2 Rock particles made by weathering of underlying rocks:
Clay: smallest and lightest particles which pack very closely together with few air spaces between.
Silt: intermediate size.
Sand: largest and heaviest particles which pack loosely together with large air spaces between.
Gravel: small stones.

3 Air: in the spaces between the soil particles. The oxygen in the soil air is needed for the respiration of plant roots, animals, and fungi and bacteria which bring about decay.

4 Water: a water film surrounds the soil particles and is taken up by plant roots. If too much water accumulates (e.g. in a clay soil) the spaces fill up with water (become waterlogged) so there is insufficient oxygen.

5 Mineral salts: such as nitrates are released by decay of dead organisms. In waterlogged soils, de-nitrifying bacteria (which can live without oxygen) break down the nitrates. Minerals may also be washed out (leached) from sandy soils.

6 Living organisms: Plant roots are mainly found in the upper layer (topsoil) which has the most humus. Scavengers (eating dead remains) such as earthworms, millipedes and springtails live in the leaf litter and topsoil. Saprophytes or decomposers (bringing about decay) such as fungi and bacteria help to recycle nutrients in the soil.

K

Comparison of soil types

Clay soil	*Loam, e.g. woodland soil*	*Sandy soil*
Small particles	Small and large	Large particles
Drainage too slow	Good drainage	Drainage too fast
Water retained in soil	Good retention of water	Water not retained in soil
Too much water	Sufficient water	Too little water
Too little oxygen	Sufficient oxygen	Sufficient oxygen
Mineral salts held to clay	Mineral salts available to plants	Mineral salts washed away

Soil fertility

(a) **Characteristics of a fertile loam soil** (i.e. a soil which is a good medium for plant and animal growth).

1 *Air* for living organisms. Need medium-sized spaces between the soil particles so there is space for soil air.

2 *Water*. Need medium-sized spaces between soil particles so that sufficient water is retained for living organisms but not so much that the soil becomes waterlogged. Also water must not drain through too fast or it will leach away mineral salts which will then not be available for plant growth.

3 *Good crumb structure:* when small soil particles have been clumped together so there are medium-sized spaces in between.

4 *Mineral nutrients*. Sufficient for healthy plant growth. Deficiencies can cause poor growth (see page 62).

Increase mineral nutrients by:

(a) Adding mulch, compost, peat or humus.

(b) Adding inorganic (chemical) fertilizers.

(c) Rotating crops (so different crops take different mineral salts from the soil from year to year).

(d) Growing legumes (beans, peas, clover) which have nodules with nitrogen-fixing bacteria. (See also page 32.)

5 *Food*. Leaf litter provides food for scavengers which in turn are eaten by other organisms. The greater the number of organisms the greater are the dead remains to be recycled.

(b) **Improving clay soil**

1 Add sand: to increase the spaces between particles so that there is more air, and water drains through more easily.

2 Add lime: this clumps the particles together and so also makes the air spaces between the particles larger. It also reduces any acidity and adds calcium to the soil.

3 Add compost and humus: these absorb some of the water so that the soil becomes less waterlogged.

4 Add fertilizers after the soil structure has been improved.

(c) **Improving sandy soil**

1 Add clay: to decrease spaces between particles, so that water does not drain through so quickly and so mineral salts are kept in the soil (not leached away).

2 Add compost and humus: these hold onto some of the water which would otherwise have drained away.

3 Add fertilizers after the soil structure has been improved.

Pollution

K

POLLUTION is the build-up of waste products so that they upset the balance in the environment. It is caused by increase in human populations, industrialization and urbanization.

Air pollution

1 **Acid rain** *Causes*: Sulphur dioxide, carbon dioxide and oxides of nitrogen produced by burning fossil fuels such as coal and oil in factories, industries and power stations. The gases dissolve in rain-water to make dilute acid solutions.
Effects: On plants – kills lichens, damages trees by causing loss of leaves, makes soil acid and reduces rate of decay.
On animals – makes fresh water acid and kills aquatic organisms, harms respiratory system of humans.
On buildings – eats away at limestone buildings.
Control: Use smokeless fuels. Build special scrubbing devices in factory chimneys to remove the harmful gases.

2 **Greenhouse effect** *Causes*: Build-up of carbon dioxide (mainly) and methane in the atmosphere causing global warming.
Effects: Hotter, wetter and more unpredictable weather.
Control: International effort to reduce greenhouse gases.

3 **Ozone holes** *Causes*: CFCs (from refrigerators, packaging etc.) react with and destroy ozone, forming ozone holes.
Effects: More ultra-violet rays to earth increasing skin cancer.
Control: Need safe substitutes for CFCs worldwide.

4 **Lead** in petrol used in cars is released into the air and can build-up (especially in children) and cause brain damage. Use of lead-free petrol in the UK has reduced lead levels in the air.

Land pollution: Pesticides

Causes: Pesticides (synthetic chemicals) such as insecticides (for killing insects) and herbicides (for killing weeds) may be on the food eaten by small animals. These animals are eaten by larger animals and so the pesticides are passed along the food chain, becoming more concentrated at each level.
Effects: Pesticides such as DDT can kill birds and cause birth defects in other animals. DDT is not now used.
Control: Use pesticides specific to the pest, that are not harmful to other organisms, that are biodegradable (break down quickly) so they do not enter the food chains. Also BIOLOGICAL CONTROL by introducing predators which just eat the pests.

Water pollution: Fertilizers

Causes: Excess fertilizers, e.g. nitrates or sewage put onto the soil, or nitrates washed out of sandy soil by rain into fresh water.

Effects: These occur in stages:

1 Increased nitrates etc. allow great increase in growth rate of algae (algal bloom).

2 As the algae die they are decomposed by bacteria which can also increase greatly in numbers.

3 The bacteria use up the oxygen in the water.

4 There is less oxygen for fish, which may die.

Note. In addition the increased nitrates in the drinking water can harm babies and in some experiments have been found to cause stomach cancer in animals.

Control:

- Only add as much fertilizer as the plants can use.
- Do not allow farmyard animal wastes (sewage) to flow into the water.
- Do not add fertilizers to sandy soils until the soil structure has been improved.

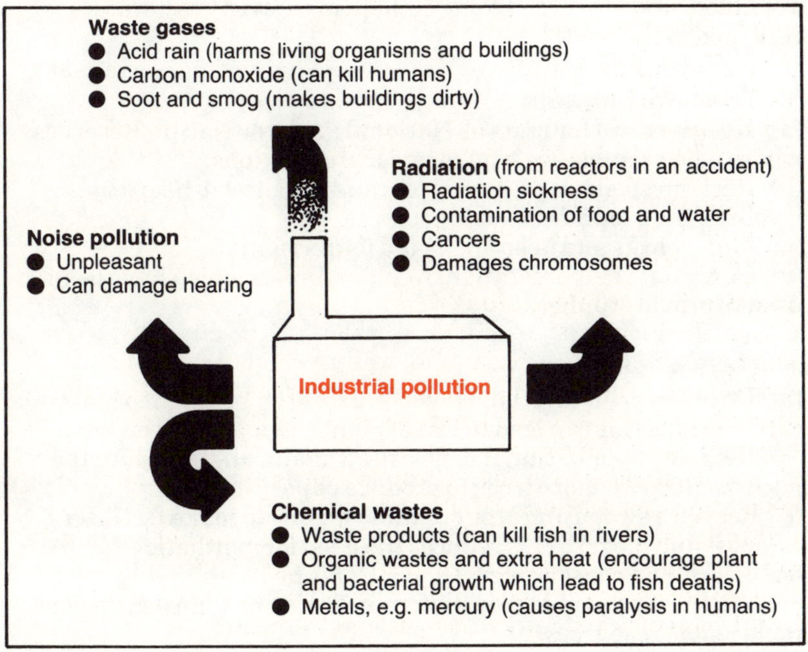

Conservation

K

CONSERVATION is the preservation and improvement of our surroundings. The things we have to do are:

1 Reduce pollution

(a) Prevent all forms of pollution (see pages 38–9).
(b) Have laws requiring factories and industries not to put waste gases into the air or harmful chemicals into the water.
(c) Use biodegradable products (that can be decomposed in nature) whenever possible. Plastics are not biodegradable.
(d) Recycle materials so that they do not accumulate.

2 Preserve species

(a) Do not remove all the hedgerows as that will kill plants and animals and the organisms that feed on them.
(b) Do not allow native organisms to be killed off by other organisms introduced from abroad.
(c) Prevent over-fishing which reduces fish populations.
(d) Protect rare and endangered species by controlling how many are killed, and also breed some of them in captivity to be later returned to the wild.

3 Preserve habitats

(a) Keep natural habitats as National Parks or Nature Reserves where wild animals are kept in natural conditions.
(b) Reclaim spoil heaps, gravel pits and colliery pit-heaps by levelling the ground and adding plants.
(c) Make parks and keep open land for recreation.
(d) Carry out re-afforestation: re-plant forests which have been used to provide timber.

4 Use resources wisely

(a) Do not over-use the land (over-grazing) by having too many cattle on it as this can lead to loss of plant cover and soil erosion.
(b) Recycle paper so that it can be used again, and so reduce the need to cut down more trees to produce paper.
(c) Recycle sewage and other organic wastes to make fertilizers, compost and biogas. This reduces the need for synthetic fertilizers, and conserves coal and oil for other uses.
(d) Whenever possible recycle materials rather than use up non-replaceable resources.

Recycling of sewage

SEWAGE is waste organic matter: urine and faeces from animals. If this is just put back onto the land or allowed to drain into fresh water it can cause problems:

1 Increase in algal and bacterial growth and subsequent death of aquatic organisms (see page 39).

2 Passing on of diseases such as dysentery, cholera and of eggs of parasitic worms in the faeces.

If sewage is treated and recycled it can be useful:

1 Sewage works separate off the water which is recycled.

2 The solid wastes can be partly decomposed by bacteria so that they are harmless and can be used as organic fertilizers.

3 Treating the sewage breaks the parasitic life cycles so infections are not passed on.

4 Sewage contains trapped energy. Sewage is decomposed by anaerobic bacteria in fermenters to produce methane (BIOGAS). This is used as fuel and saves on the use of fossil fuels.

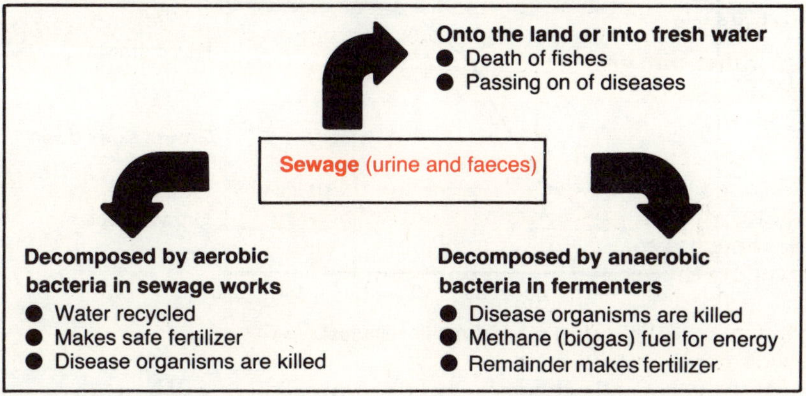

Summary

1 Humans need to use the land, but they can cause problems through deforestation, soil erosion, monoculture, pesticides and pollution.

2 The main causes of air, land and water pollution are acid rain, pesticides and excess fertilizers and sewage.

3 Conservation depends upon reducing pollution, preserving species and habitats and using resources wisely.

7 Controlling population growth

Aims of the chapter

After studying this chapter you should be able to:

K

1 Describe the stages of population growth.
2 Explain how various factors affect population growth.
3 Discuss factors affecting human population growth.

Stages of population growth

If we consider a simple population of yeast in a sugar solution, the yeast will grow and multiply because it has plenty of food. The number of yeast cells will increase.

We can draw a graph, as shown below, with the number of yeast cells on the upright (vertical or *y* axis), and the time along the horizontal (or *x* axis). We record the number of yeast cells every 20 minutes (shown by dots) and then join up the ⊙ marks to make a curve. This is called a GROWTH CURVE.

K

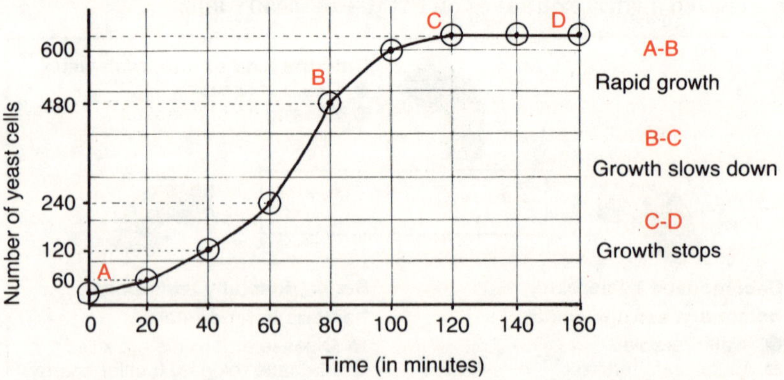

If the yeast cells divide every 20 minutes, there will be a doubling of the population every 20 minutes. This is the stage of rapid or **exponential growth**. It is shown by the first, steep part of the curve **A**–**B**.

But after a while the food will begin to run out, waste products may have accumulated, and the container may be too small, and so the growth rate slows down. The growth curve becomes less steep, as shown by the part **B**–**C**.

If conditions continue to become difficult, the yeast cells stop dividing, so there is no increase in numbers. So the curve becomes flat, as shown by the part **C**–**D**.

Factors which affect population growth

1 Food

If there is plenty of food, population growth will be rapid. For a herbivore (primary consumer) this means plenty of the kinds of food that it eats. For a carnivore (predator) it means sufficient prey.

Producer → Herbivore → Carnivore
(Prey) (Predator)

The predator increases in numbers by eating the prey. So the prey is reduced in numbers. But then the predator will not have enough food and so some will die. At that time the prey will be able to increase in numbers. And so on.

2 Poisonous waste products

When yeast grows in sugar solution, it produces alcohol (ethanol) which eventually kills it. Other populations also produce waste products which may kill them.

3 Overcrowding

As populations increase, there is less space for further growth. In nature, when populations are crowded there is often a slowing down in population growth.

4 Competition

If there is overcrowding, then some of the organisms will fight with the others for food, or for mates. Competition may cause death and reduce numbers.

5 Disease

If there is insufficient food the organisms may be unhealthy. They may also be overcrowded and weakened from competition. In these conditions diseases can spread rapidly and kill some of the population.

6 Humans

K Humans control other populations, for example by the use of pesticides and biological control. Humans also develop new strains of organisms and decide which crops to grow.

K All these factors which control how large a population may grow are called LIMITING FACTORS.

Human population growth

Food. The increase in human population size and the uneven distribution of food means that some people have food mountains while others face malnutrition (unhealthy diet) or famine and starvation (death from insufficient food).

Poisonous waste products. For the human population these would be the various harmful pollutants which we produce.

Overcrowding. In some cities of the world this leads to poor housing and unemployment which can affect health, and allows **disease** to spread faster.

Competition between individuals and nations leads to the 'survival of the fittest'. Conflicts can cause wars.

K In addition humans exercise various methods of BIRTH CONTROL, and so keep down their population numbers.

1 **Barrier (mechanical) methods** stop egg and sperm meeting:
(a) Condom used by the man (also protects against the spread of sexually transmitted diseases (STDs)).
(b) Diaphragm (cap) used by the woman to cover the entrance to the uterus.

2 **Chemical methods**
(a) Contraceptive pills.
(b) Spermicides (creams or jellies) inserted into the vagina before intercourse to kill sperm.
Note. It is more effective to use spermicides together with a condom or cap than to use either method separately.

3 **Surgical methods** remove the pathway for sperm or eggs.
(a) Vasectomy in men: cutting of the sperm ducts.
(b) Tubal ligation in women: cutting of Fallopian tubes.

4 **Natural methods.** The least effective.
The 'rhythm method': having intercourse only on the so-called 'safe days' around the time of the menstrual period.

Summary

1 Factors limiting population growth are food, poisonous wastes, overcrowding, competition and disease.
2 The human population is controlled by the same limiting factors that control other populations. But we can also think about the situation and take appropriate action.
3 Birth control methods control human population growth.

Aims of the chapter

After studying this chapter you should be able to:

1 Describe the relationship between a parasite and its host.
2 Name a parasite and describe how it is controlled.
3 Give an account of a sexually transmitted disease (STD).
4 Describe how we can be protected against disease.

Interrelationships between organisms

A PARASITE is an organism which feeds on another living organism (the host). The parasite benefits, but it harms the host. Disease-causing parasites are called PATHOGENS.

A HOST is a living organism which is used as a food source by a parasite. It is harmed in some way.

A VECTOR is an organism which carries a parasite from one host to another. For example, the mosquito which transfers malarial parasites from one human to another.

EXTERNAL or Ectoparasites live on the outside of their hosts, e.g. lice, and mistletoe.

INTERNAL or Endoparasites live inside their hosts.

Examples of endoparasites

Group	*Diseases caused*
Viruses	Flu; Herpes and AIDS (STDs)
Bacteria	Gonorrhoea and Syphilis (STDs); dental decay, Tuberculosis (TB), food poisoning
Protozoa	Malaria
Fungi	Athlete's foot, Potato blight

Very small organisms that can only be seen with the microscope are called MICRO-ORGANISMS or MICROBES, e.g. viruses, bacteria, protozoa and some fungi. Microbes can be:
1 Disease producing (pathogenic): see table above.
2 Useful, e.g. nitrogen-fixing bacteria in the nodules of leguminous plants where they are SYMBIOTIC (both the bacteria and the plant benefit from the relationship).
e.g. bacteria and fungi in biotechnology (see page 29).

Some examples of parasites

The head louse: an ectoparasite
Adaptations

Flattened shape – so lies close to skin
Claws on legs – to hold onto hairs
Biting mouthparts – to suck blood for food

Significance: Bites cause itching and can spread disease-causing bacteria and viruses into the blood.
Control: Wash hair with a special shampoo which kills adults and egg cases ('nits') which can then be combed out.

The bacterium causing gonorrhoea (an STD)
Transmission: During sexual intercourse with an infected person.
Symptoms and effects: 1–2 weeks from infection
Man – liquid (pus) from penis. Discomfort on urinating.
Woman – increase in vaginal discharge or no visible symptoms.
Prevention: Man should use a condom. Sexual intercourse with only one person who is not infected.
Treatment: A course of antibiotics (such as penicillin).
Dangers: STDs may be passed on before a person is aware that they have the disease. If the full course of antibiotics is not taken then the bacteria can develop resistance.

Malarial parasite: a protozoan causing malaria.
Life cycle: The malarial parasite spends part of its life in the Anopheles mosquito, and part in the liver and red blood cells of humans. It increases in numbers in both organisms.
Transmission: If an infected mosquito bites a human it puts saliva (containing parasites) into the blood. These parasites spend most of their time reproducing in the red cells. When the red cells break up, the person has a high fever.
Control in humans
If untreated, malaria can kill. Drugs, based on quinine, can kill most kinds of the parasite. It is better to avoid being bitten by wearing insect-repellent cream and sleeping under mosquito nets when travelling in areas with malaria.
Control by killing mosquitoes (the vector)
- Spray oil on water to kill mosquito larvae and pupae.
- Spray insecticides to kill adult mosquitoes.
- Drain swamps and other breeding places to reduce mosquitoes.

Protection against disease

(a) **Keep yourself clean.** Any germs (pathogenic micro-organisms) on the skin could get in through small cracks. So it is important to wash the skin, and to cover large wounds.

(b) **Keep food clean.** Pots and pans should be kept clean so we do not get food poisoning. Food should be kept refrigerated, and not eaten if it smells bad or has been kept too long.

(c) **Avoid certain foods** if you are advised to do so. For example, some kinds of pâté and cheese are liable to infection from *Listeria*. When buying pre-cooked convenience foods, make sure to heat them sufficiently in an oven before eating them.

(d) **Use antiseptics** e.g. antiseptic mouthwashes and creams, and toilet cleaners which prevent growth of bacteria. Discs of filter paper soaked in antiseptic and placed on a dish containing bacteria will kill the bacteria close by.

K (e) **Use antibiotics** e.g. penicillin as pills or injections. These are drugs used to kill bacteria when we have become infected. They are prescribed by a doctor. Their action can be shown by antibiotic discs on dishes of bacteria.

K (f) **Be immunized** (protected against disease) by producing your own antibodies or receiving ready-made antibodies against diseases caused by viruses and bacteria.

(i) *Producing antibodies* – ACTIVE immunity
(a) Our white blood cells (lymphocytes) produce antibodies in response to invading organisms which have special chemicals (antigens) on their surface. Each kind of microbe causes the production of a special antibody which can later kill any other invading microbes of the same kind.
(b) We can be vaccinated by injection of dead or weakened viruses or bacteria. We do not become ill, but our white blood cells make the right antibodies against future infection.
(*Note*. People with AIDS cannot make sufficient antibodies and so they may die from many kinds of disease.)

(ii) *Receiving ready-made antibodies* – PASSIVE immunity
(a) Antibodies from the mother can pass through the placenta and through breast milk to protect the baby.
(b) Injections of antibodies for protection, e.g. against cholera, can be given by a doctor.
(*Note*. Passive immunity is not long-lasting. For better protection the person has to make their own antibodies.)

Questions on Chapters 3–8

For Questions **8–23**, each group of questions has five possible answers (A),(B),(C),(D) or (E). For each question choose the best response. A response may be used once, more than once or not at all.

Questions 8–11

(A) Habitat (B) Population (C) Community
(D) Environment (E) Ecosystem

8 The conditions found in a certain place. ____
9 A group made up of several species living in a certain place. ____
10 The place where an organism lives. ____
11 A group of organisms that can interbreed. ____

Questions 12–15

(A) Line transect (B) Quadrat (C) Plankton net
(D) Tullgren funnel (E) Butterfly net

12 Used to collect aquatic organisms. ____
13 Used to collect flying insects. ____
14 Used to find the density of a population. ____
15 Used to find the percentage cover of a population. ____

Questions 16–19

(A) Producers (B) Decomposers (C) Primary consumers
(D) Secondary consumers (E) Tertiary consumers

16 The organisms with the least total biomass. ____
17 Organisms which do not depend on other organisms for food. ____
18 Organisms which eat and are eaten by animals. ____
19 Organisms which recycle nutrients. ____

Questions 20–23

(A) Plastics (B) Paper (C) Sewage
(D) Fossil fuels (E) Compost

20 Anaerobic decomposition of which item produces methane (biogas)? ____
21 Which item is LEAST biodegradable? ____
22 Which item is MOST directly derived from plants? ____
23 Which item would it be best to add to the soil? ____

24 The figure below shows part of a food web in the garden.

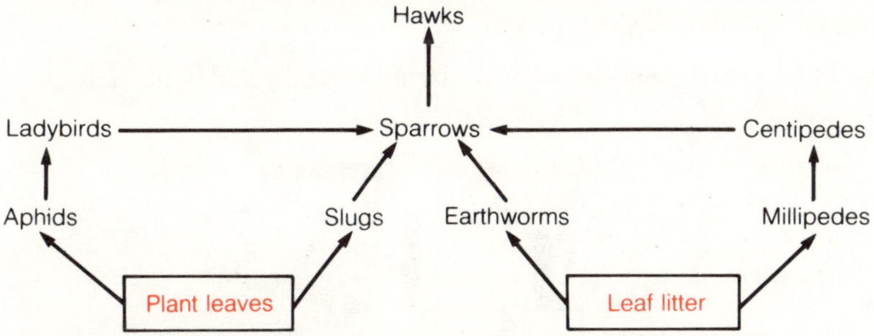

Using the information in the food web above:
(a) Write out a food chain which has five organisms in it.
(b) Name a primary consumer.
(c) If all the ladybirds died
  (i) What would be the effect on the aphids?
  (ii) What would be the effect on the sparrows?

25 Which of the following statements concerning the pyramids of biomass and numbers is CORRECT?
(A) Biomass increases at higher trophic levels.
(B) Energy is lost at each trophic level.
(C) More organisms can be supported at higher trophic levels.
(D) There are fewer organisms at the lower trophic levels.

26 Shown below are four food chains involving humans.
(A) Baked beans → humans
(B) Grass → cows → humans
(C) Plankton → small fish → herrings → humans
(D) Grain → chickens → humans

In which case would the greatest amount of energy, as trapped during photosynthesis, be provided to the human? (A), (B), (C) or (D)?

27 (a) Name an air pollutant, and describe the damage that it does to plants or animals.
(b) Plants need nitrates to make proteins. But adding nitrate fertilizers to the soil can lead to the killing of fish. Explain why this is so.

28 List three ways in which man has affected the environment. In each case write ONE sentence to describe how plants or animals have been affected.

29 The figure below shows how nitrogen is recycled in nature.

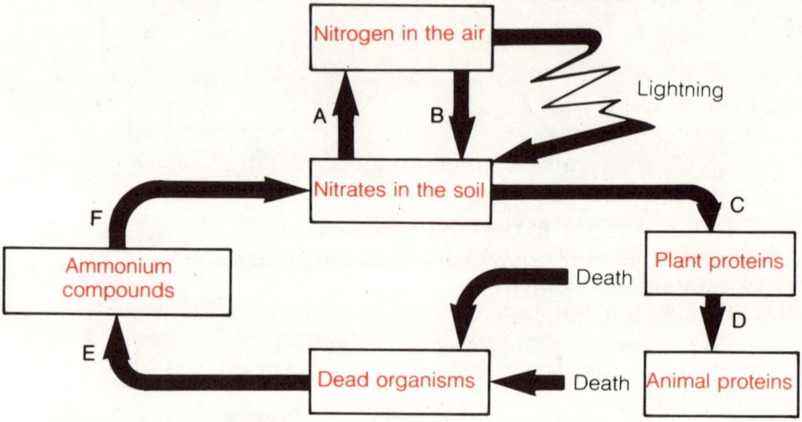

(a) Name the processes **C** and **D**.
(b) The arrows labelled **A,B,E** and **F** represent four kinds of bacteria involved in the nitrogen cycle.
Which of these bacteria:

(i) Increase the amount of nitrates available for plants? ____ and ____
(ii) Can live in the nodules of leguminous plants? ____
(iii) Decrease the amount of nitrates in waterlogged soils? ____

(c) Which of the processes **A–F** can also be carried out by saprophytic fungi? ____

30 (a) List four factors which control population growth.
(b) List an *additional* way in which the human population is reduced.
(c) List an *additional* way in which the human population is increased.

31 Describe what is meant by the following words:
(a) Parasite (b) Antibiotic (c) Antibodies

Now turn to page 142 to check your answers 8–31.

9 Cells as building blocks

Aims of the chapter

After studying this chapter you should be able to:

1 Describe the parts of a cell.
2 Distinguish between plant and animal cells.
3 Define and give examples of cells, tissues and organs.

Parts of a cell (as seen under a light microscope)

Cytoplasm. Jelly-like substance (about 70 per cent water) where chemical reactions occur and food is stored.
Nucleus. Controls development and reproduction. Contains the chromosomes which pass on characteristics.
Protoplasm. Consists of cytoplasm and nucleus.
Cell membrane. Surrounds the cytoplasm and controls what enters and leaves the cell.
Mitochondria. Small bodies where aerobic respiration occurs.
Plant cells only
Chloroplasts. Small green bodies containing chlorophyll needed for photosynthesis.
Cell wall. Cellulose wall outside the cell membrane. Makes the plant cell keep its shape.
Large vacuoles. Spaces containing liquid (cell sap).

Plant and animal cells: similarities and differences

Similarities

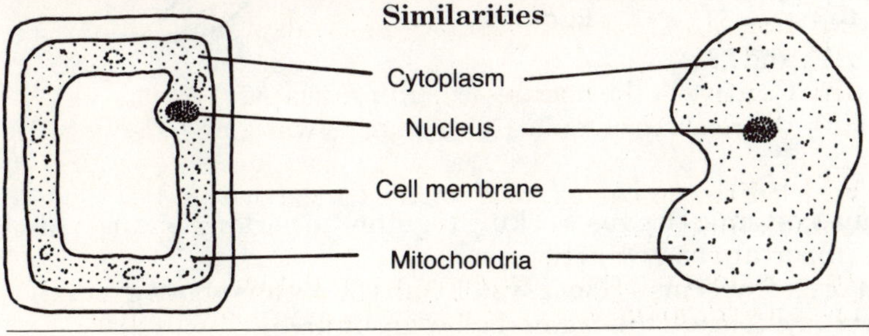

Differences

Plant cells	*Animal cells*
1 Chlorophyll in chloroplasts	No chlorophyll or chloroplasts
2 Cellulose cell wall	No cellulose cell wall
3 Large vacuoles	Vacuoles small or absent

K

Cells. Basic units of living things. Protoplasm (cytoplasm plus nucleus) surrounded by the cell membrane.
Plant cells – Simple (e.g. cells of onion epidermis) or specialized (e.g. root hair cells and cells in a leaf).
Animal cells – Simple (e.g. cheek cells) or specialized (e.g. egg and sperm cells, blood cells).

K

Tissue: Made from similar cells grouped together and performing a particular function.
Plant tissues – Protective (epidermis), Transporting (xylem and phloem), Photosynthetic (palisade mesophyll in the leaf).
Animal tissues – Protective (epidermis), Transporting (blood), Supporting (bone), Communicating (nerves).

K

Organs: Different tissues working together to perform one or more special functions.
Plant organs – Stem, leaf, roots, flowers and seeds.
Animal organs – The stomach, liver, eye.

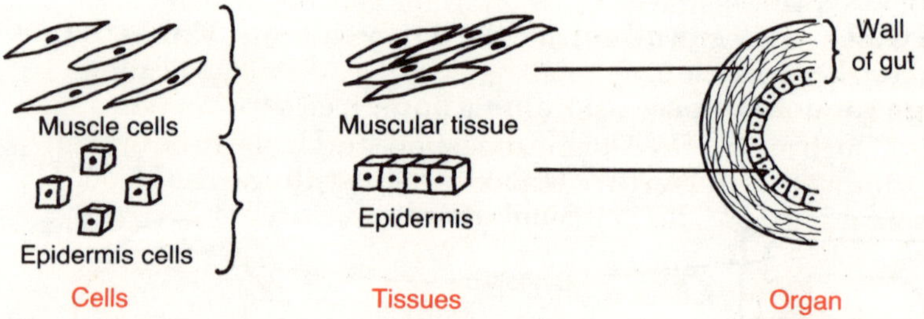

Organ systems: Organs working together to perform one or more important body functions.
Plant organ systems – Shoot system (all the organs above ground) and root system (all the organs below ground).
Animal organ systems – Digestive system, Respiratory system, Circulatory (transport) system, Excretory system, Nervous system, Reproductive system.

Organism: Whole individual. Can be very simple, and consist of only one cell (e.g. *Amoeba*) or be complex (e.g. flowering plant and mammal) when it will contain many organ systems.

Aims of the chapter

After studying this chapter you should be able to:

1 Define and explain diffusion and give examples.
2 Define and explain osmosis and give examples.

Diffusion

Diffusion means movement. Gases and liquids have particles which are continually moving in a random way. The particles tend to spread themselves out from a place where there are a lot of them (high concentration) to a place where there are few (low concentration). By chance some particles will go 'backwards', but there is an overall or **net** movement from high to lower concentration.

DIFFUSION is the net movement of particles from a region of higher concentration to a region of lower concentration. The particles are said to travel along a DIFFUSION or CONCENTRATION gradient.

In living organisms the cell membrane controls which substances enter and leave the cells. The membrane can be thought of as having very small holes or pores which will allow through some particles (such as glucose and water), but not others (such as starch). During digestion starch has to be broken down into glucose before it can pass into the villi and the blood vessels. A model of this is shown below.

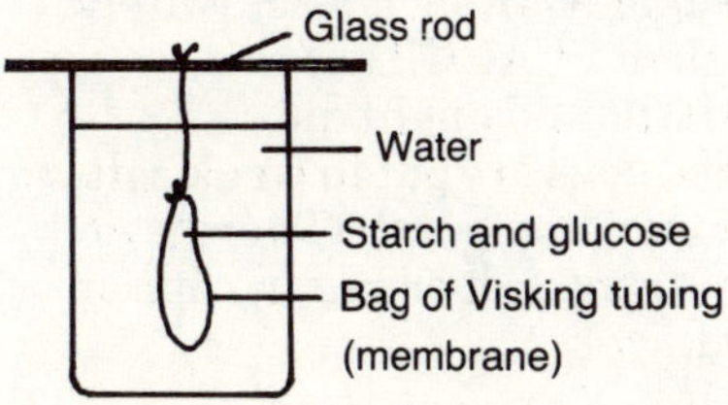

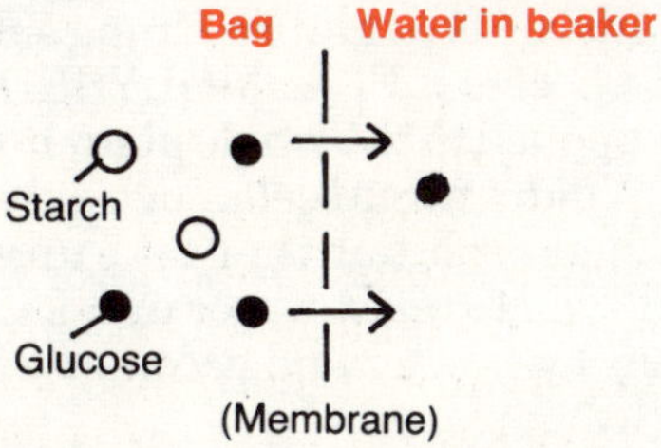

Diffusion occurs wherever gases are moving (e.g. respiration and photosynthesis) or food is moving (e.g. into villi).

Active transport occurs when particles move against their diffusion or concentration gradient. This requires energy (e.g. uptake of mineral salt ions by root hair cells).

Osmosis

Osmosis is a kind of diffusion. Osmosis is only concerned with the movement of water. It occurs when there is a membrane (selectively permeable) which keeps back the dissolved particles (solute) and only lets the water (solvent) through.

OSMOSIS is the net movement of water through a selectively permeable membrane from a region where it is in higher concentration to a region where it is in lower concentration.

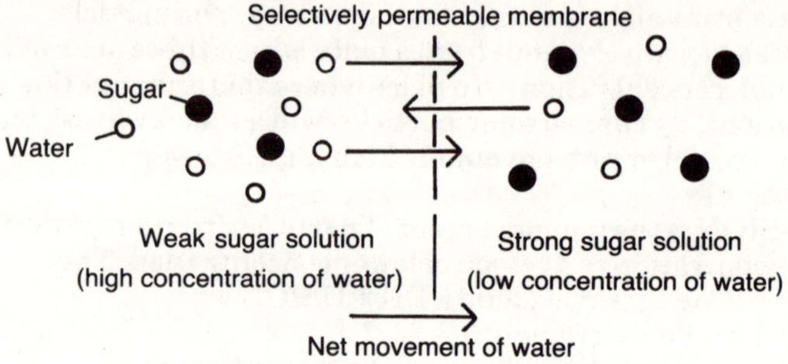

Osmosis occurs wherever water is taken up (e.g. by roots) or is transferred from cell to cell.

Plants can take up water by osmosis from a salt or sugar solution which is less concentrated than their cell sap. They become swollen with water or TURGID and this helps to support the plant. Their cellulose cell wall stops more water entering. If a plant cell is put into a strong salt or sugar solution it loses water and becomes PLASMOLYSED and then FLACCID. If this happened to the whole plant it would wilt and might die.

When **animal** cells such as red blood cells are put in weak salt or sugar solution they take up water and may burst. (There is no cell wall to resist water uptake.) In a strong salt or sugar solution they lose water and become wrinkled.

Summary

1 Diffusion is the net movement of particles from a region of higher concentration to a region of lower concentration (i.e. along a concentration or diffusion gradient).
2 Osmosis is the diffusion of water.
3 Active transport involves the expenditure of energy.

11 How enzymes work

Aims of the chapter

After studying this chapter you should be able to:

1 Describe the characteristics of enzymes.
2 Give examples of where amylases are found.
3 Describe experiments showing the effects of temperature and pH on amylase activity.
4 Give examples of the action of other enzymes.

Characteristics of enzymes

Enzymes are biological CATALYSTS which speed up reactions in living organisms. They may help in building up complex molecules or breaking them down.

Property	*Notes*
1 Not used up in reactions.	Act as catalysts. Speed up reactions but are not themselves changed.
2 Specific.	Catalyse specific reactions only, e.g. amylases act on starch.
3 Action increased by temperature.	The higher the temperature, the greater the enzyme activity.
4 Made of protein.	When heated above 40°C their activity is reduced; the protein is damaged.
5 Sensitive to pH.	Each enzyme works best at a certain pH level.

Amylases and starch digestion

Amylases are enzymes which catalyse the breakdown of starch to soluble sugars.

There are different kinds of amylase:

1 Salivary amylase which is produced in the salivary glands of mammals and which acts in the mouth to help begin the digestion of some starches.
2 Amylase produced in the pancreas which helps with the digestion of all kinds of starch to soluble sugars.
3 Amylases are found in the seeds of plants and in tubers such as potatoes. They help to digest the starch food stores and make them into soluble sugars which can then be transported and used for growth. An amylase in seeds is diastase.
4 Amylases secreted by saprophytic fungi and bacteria onto dead organisms which makes their starches soluble; part of the decay process.

Plant amylase (diastase) and the effect of temperature

Amylase catalyses the change of starch to soluble sugars. So if the enzyme acts, the starch disappears. The experiment is set up at different temperatures.

To make it a fair test, all the other conditions or VARIABLES, except for the temperature, have to be the same. This is called a K CONTROLLED EXPERIMENT. The CONTROL has all the variables we think are important, and each set-up is different from this in only one variable:

1 Make a mixture of crushed dry biscuit (starch) with water.
2 Put equal quantities into five test tubes.
3 Get the starch mixtures to the required temperatures, e.g.:
   (i) Put one in a beaker with ice (about 0°C).
   (ii) Leave one at room temperature (about 15°C).
   (iii) Put two into water baths at different temperatures (about 25°C and 35°C).
4 At the same time add the same amount of diastase (from crushed germinating seeds) to the four test tubes above.
5 Take out samples and test them with iodine solution at 2 minute intervals. The enzyme will work fastest in the test tube at 35°C (the OPTIMAL temperature).
6 Boil the same amount of diastase for 5 minutes, then add it to the fifth test tube. When tested with iodine solution there is no change. The enzyme has been denatured by boiling.

Plant amylase (diastase) and the effect of pH

1 Make a mixture of crushed dry biscuit with water.
2 Put equal quantities into three test tubes.
3 (i) Test tube 1 contains dilute hydrochloric acid ($<$pH7)
   (ii) Test tube 2 contains dilute sodium hydroxide ($>$pH7).
   (iii) Test tube 3 contains distilled water (neutral pH7).
4 Add equal amounts of diastase to the test tubes.
5 After the same amount of time (e.g. 5 minutes) test drops of liquid from the test tubes with iodine solution.
6 Test tubes 1 and 2, iodine goes blue-black to show starch is still present (i.e. starch has not been digested).

Test tube 3, iodine solution remains straw coloured, so starch has been digested.

7 Conclusion: Diastase works best in neutral conditions. This is called its OPTIMAL pH.

Other important enzymes

1 Building up of starch. On the previous page you saw how starch could be broken down. But another enzyme can build up glucose into starch. Enzymes are SPECIFIC (restricted) as to which reaction they catalyse. This means they only act on certain substances (called SUBSTRATES).

2 Digestive enzymes. These catalyse the breakdown of complex insoluble substances into simpler soluble ones:
Starch — amylases → Simple sugars
Fats — lipases → Fatty acids and glycerol
Proteins — proteases → Amino acids

3 'Biological' washing powders. These contain enzymes which can digest biological stains which contain proteins, such as blood, gravy, egg yolk, perspiration, chocolate and milk. The instructions for the use of biological powders (each has a reason related to the characteristics of enzymes) are given below:
(a) Soak the dirty clothes before the main wash. (This is because the washing cycle temperature may destroy the enzyme activity.)
(b) Soak in hand-hot water. (This is a high enough temperature to allow greater enzyme activity without destroying the enzyme.)
(c) Do not soak silk, wool or leather. (These contain protein and may themselves be digested.)
(d) Always rinse and dry the hands after use. (This is to remove any enzymes which could harm the skin.)

If you are asked to compare washing powders which are 'biological' (with enzymes) and 'non-biological' (without enzymes), make sure it is a fair test (a controlled experiment), for example:
(i) Use the same amount of the same kind of cloth with the same stain.
(ii) Set up pieces of both samples at a range of the same temperatures with the same amount of powder.
(iii) Decide how you will judge when they are clean.

Summary

1 Enzymes act as catalysts for specific reactions. They are made of proteins and are sensitive to temperature and pH.
2 Digestive enzymes are amylases, lipases and proteases.
3 Biological washing powders contain enzymes.

12 How photosynthesis traps energy

Aims of the chapter

After studying this chapter you should be able to:

K
1 Describe experiments which show that carbon dioxide, light and chlorophyll are necessary for photosynthesis.
2 Describe experiments which show that food and oxygen are produced during photosynthesis.
3 List ways in which leaf structure helps photosynthesis.

Photosynthesis experiments

K

In photosynthesis experiments:
(i) DE-STARCHED plants are used. These have been left in the dark for 24 hours (e.g. in black plastic bags) so that their starch stores will have been removed from the leaves.
(ii) Each experiment is set up with a control.
(iii) After the experiment leaves are tested for starch (page 60).

1 **Carbon dioxide is necessary for photosynthesis:** Two leaves of a potted plant are enclosed in boiling tubes. In one tube are placed some sodium hydroxide pellets to absorb carbon dioxide (I). The other tube serves as a control which still has a supply of carbon dioxide (II).

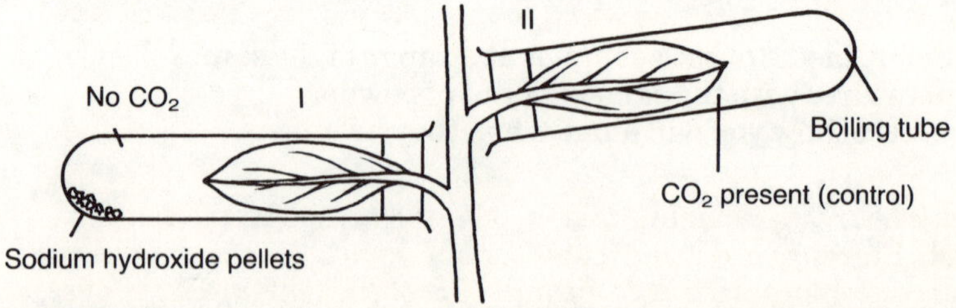

After 3 hours in sunlight the leaves are tested for starch. Only the leaf (II) which was supplied with carbon dioxide has carried out photosynthesis.

2 Light is necessary for photosynthesis: During photosynthesis light energy is converted into chemical energy stored in sugars (and starch). Simple substances (carbon dioxide and water) are combined into complex energy-containing ones.

The leaf of a de-starched plant is partly covered by aluminium foil (a). The uncovered part serves as the control.

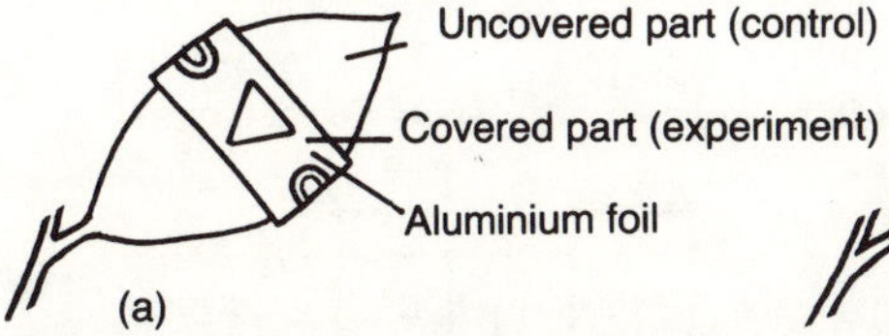

(a)

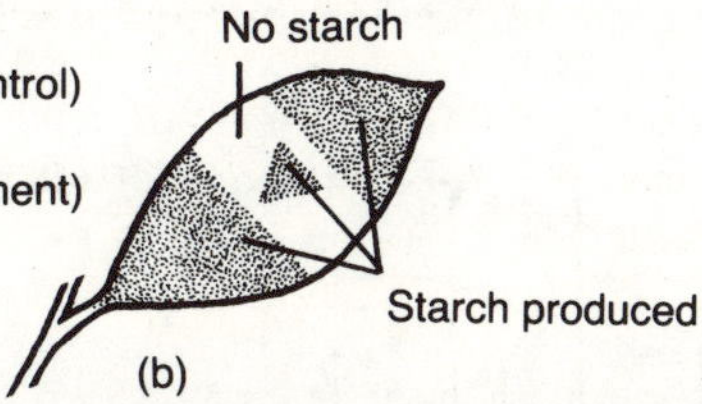

(b)

The plant is left in the light for 3 hours. The leaf is removed and tested for starch. Starch is produced only in the part of the leaf which received light (see (b) above).

Note. Increasing the amount of carbon dioxide or of light, e.g. in a greenhouse, increases the rate of photosynthesis. Smoke pollution decreases photosynthesis.

3 Chlorophyll is necessary for photosynthesis: Chlorophyll absorbs the light energy used in photosynthesis. A de-starched plant with variegated leaves is used. The leaves are partly green (with chlorophyll) and partly white (without chlorophyll) (see (a) below).

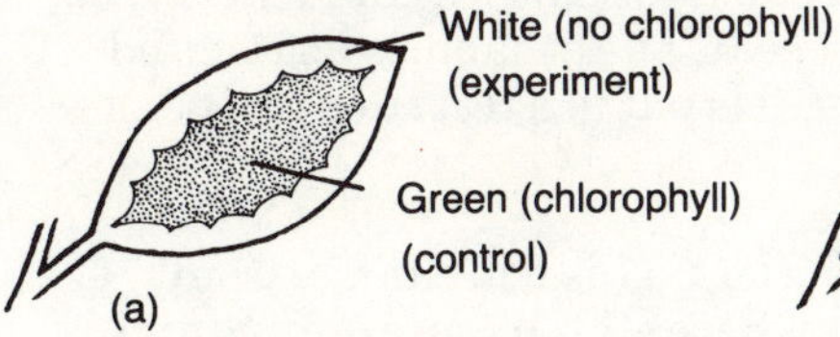

(a)

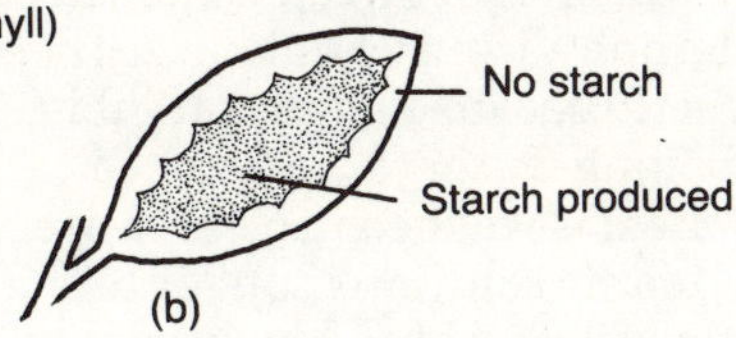

(b)

After 3 hours in the light, the leaf is removed and tested for starch. Starch is produced (that is, photosynthesis occurs) only in the green (chlorophyll-containing) part of the leaf (see (b) above).

Note. Photosynthesis is controlled by enzymes which work faster at higher temperatures. That is why greenhouses are kept warm to increase the rate of photosynthesis.

Products of photosynthesis

1 Oxygen is released during photosynthesis: A water plant such as *Elodea* (Canadian pondweed) is put in water which contains plenty of carbon dioxide, and left in the light (a) below. Bubbles of gas are produced (b). When tested the gas re-lights a glowing splint showing it contains oxygen. Increasing the light intensity increases the rate at which oxygen bubbles are produced.

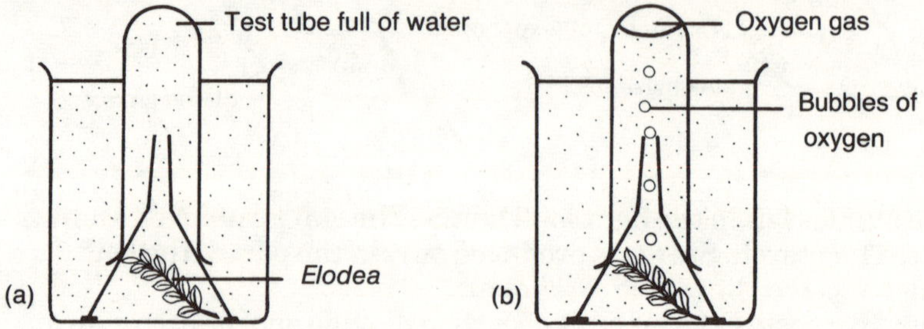

2 Sugars are produced during photosynthesis which are quickly changed to **starch**. Sugar production can be shown, e.g. in leaves of iris. Starch production is shown in most leaves, e.g. geranium. The test for starch is given below.

K The STARCH TEST

1 Leaf into boiling water to kill and soften it.
2 Leaf into boiling alcohol (ethanol) to remove (EXTRACT) the chlorophyll and de-colourize the leaf. The ethanol is not heated directly because it is flammable. It is put in a test tube in a beaker of boiling water (a water bath).
3 Leaf is washed in clean water.
4 Iodine solution is put onto the leaf. This will change from yellowish to a blue-black colour where starch is present. Starch production shows photosynthesis has occurred.

Leaf structure is adapted for photosynthesis

Leaves are held up by the leaf stalk (petiole) so that light falls onto them. They are thin and flat so that there is a large surface area through which light energy can enter. The mid-rib and veins support the leaf blade (lamina). The adaptations of the internal leaf structure are shown opposite.

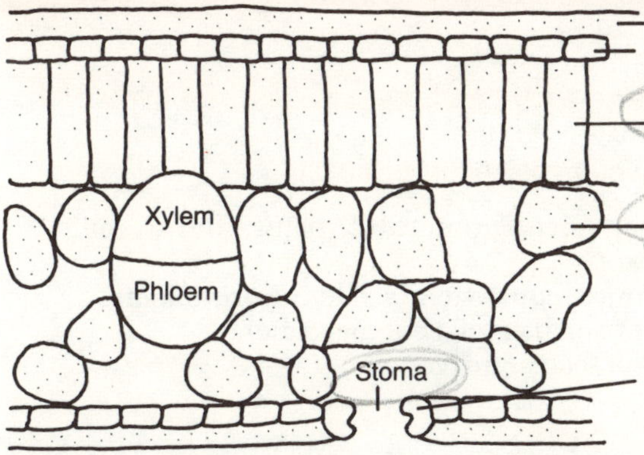

1 Epidermis is transparent – lets light through.
2 Palisade mesophyll – contains most chloroplasts which contain chlorophyll. Cells are in a single layer and chloroplasts can easily trap energy for photosynthesis.
3 Xylem with xylem vessels which bring water to the leaf for photosynthesis.
4 Phloem with sieve tubes which take food away from the leaf to all parts of the plant.
5 Spongy mesophyll – round, loosely packed cells with many air spaces in between, which allow for the diffusion of carbon dioxide, oxygen and water vapour.
6 Pores (stomata), mostly in the lower epidermis, allow the entry of carbon dioxide for photosynthesis and the removal of oxygen. The stomata are enclosed by the guard cells which are open in the daytime, and closed at night when photosynthesis is not occurring.
7 Cuticle covers the epidermis except for the stomata. This restricts the loss of water.

Summary

1 All organisms depend upon energy trapped by photosynthesis.
2 It requires carbon dioxide, water, light and chlorophyll.
3 Oxygen released during photosynthesis is the only source of oxygen for the respiration of living things.
4 Sugars produced during photosynthesis are changed to starch and cellulose, to fats, and to amino acids (with the nitrates from the soil water) to make proteins.

13 Substances needed for healthy growth

Aims of the chapter

After studying this chapter you should be able to:

K

1. Describe the importance of minerals to plants and animals.
2. Carry out food tests.
3. Describe the chemical constituents of living organisms.
4. Describe the parts and uses of a balanced diet.
5. Evaluate the use of food additives.

Mineral requirements in plants

K

In plants simple chemicals are taken in (carbon dioxide, water and minerals) and built up into carbohydrates, fats and proteins. The importance of different minerals is shown by growing seedlings (with their cotyledons removed) in liquids which lack certain minerals (CULTURE SOLUTIONS).

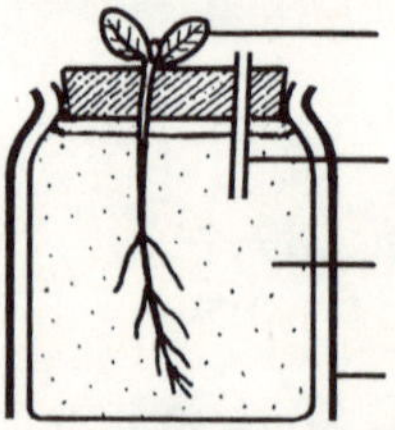

Seedlings of the same age and size are put into 4 different solutions, e.g. without nitrogen, calcium, iron and magnesium. A fifth one with all the minerals is the control.

Air is added for the respiration of the roots. The dark cover prevents algae growing.

Without *nitrates*: very stunted growth as nitrogen is needed to build all proteins. (See also the role of fertilizers, page 33.)
Without *calcium*: poor growth, as needed for cell walls.
Without *iron*: yellow (chlorosis) as iron is needed in the making of chlorophyll (though it is not part of the molecule).
Without *magnesium*: yellow. It is part of chlorophyll molecule.
Control with all minerals: Healthy growth.

Mineral requirements in animals

Many minerals are necessary in small amounts, for example:
Iron: Is part of haemoglobin in the red cells which transport oxygen. Lack of iron can cause anaemia (weakness and tiredness due to insufficient haemoglobin). Sources: liver, spinach.
Calcium: Used in building bone, so important for the skeleton and for healthy teeth. Lack causes poor bones and teeth. Sources: milk, cheese.

Food tests

Starch test, e.g. potato **(Iodine solution)**
A few drops of iodine solution are added to the food to be tested. Iodine stains the starch grains blue-black. When testing for starch in a leaf the chlorophyll is first removed (see page 60), so it does not mask the colour change.

Simple **sugars**, e.g. glucose **(Benedict's reagent)**. Mix a little of the food with water, then proceed as below:

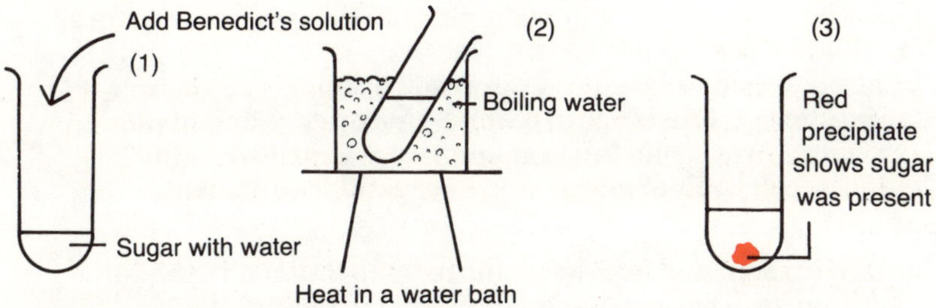

Fat, e.g. margarine **(Grease spot test)**. Rub a little of the food onto a piece of thin paper. If the food contains fat it will leave a greasy translucent spot (which allows only some light through).

Protein, e.g. egg white **(Biüret test)**. Mix a little of the food with water, then proceed as below:

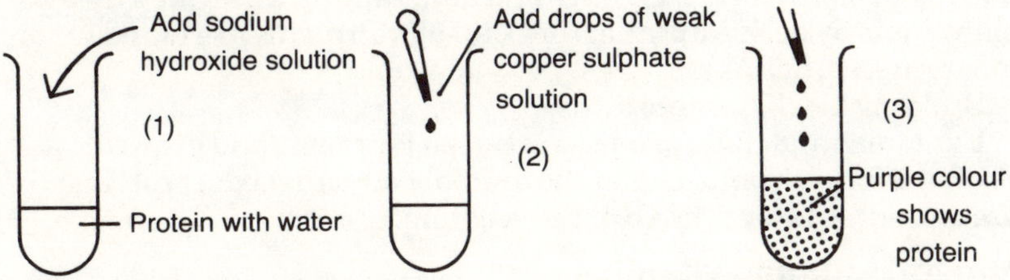

Vitamin C, e.g. lemon **(DCPIP test)**:

1 Put one drop of blue DCPIP solution on a white tile.
2 Add the juice to be tested drop by drop, and stir.
3 If vitamin C is present, the DCPIP goes colourless.

Chemical constituents of living organisms

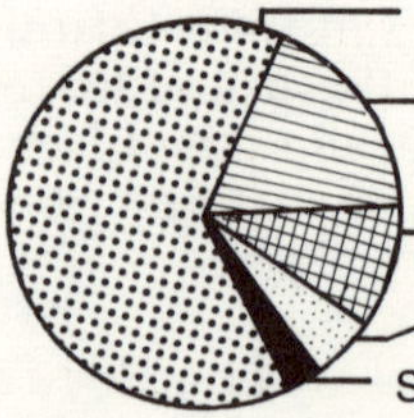

Water

Contains hydrogen and oxygen (H_2O). Living organisms are mostly water. This is why water is so important for life (see page 34).

Carbohydrates

Contain carbon, and hydrogen and oxygen (in the same proportion as in water). For example glucose, sucrose and starch.

Building blocks: glucose.

Examples and uses: glucose for respiration (energy source).
Starch: storage substance in plants, e.g. potatoes and bread.
Glycogen: storage substance in animals, e.g. in liver.
Cellulose: cell walls of plants, e.g. in vegetables and fruit.

Fats

Contain carbon, and hydrogen and oxygen (but not in the same proportion as in water). For example, in butter and oils.

Building blocks: fatty acids and glycerol.

Examples and uses: saturated fats (mostly animal fats).
Unsaturated fats (mostly plant fats and oils).
Fats are stored in seeds, and under the skin in mammals for insulation. Fats are respired for an energy source.

Proteins

Contain carbon, hydrogen and oxygen, and also nitrogen. (Nitrogen source is the nitrates and ammonium salts taken up by plant roots by active uptake from the soil.) Proteins are found in meat, cheese, beans and eggs.

Building blocks: amino acids.

Examples and uses: proteins are used for repair and growth.
Proteins form the muscles of the animal body. In seeds, proteins are stored for later growth of the seedling.

Mineral salts and vitamins

These are needed in small amounts for health.

Mineral salts often become part of other chemical constituents, e.g. nitrates become part of proteins, and magnesium becomes part of the chlorophyll molecule. Vitamins often act like enzymes to speed up chemical reactions in the body. They allow other compounds to be used properly in the body.

A balanced diet

Plants are AUTOTROPHIC ('self-feeding'). They can make their own food from simple chemicals (carbon dioxide, water and mineral salts) in the presence of light and chlorophyll.

Animals are HETEROTROPHIC ('other feeding'). They cannot make their own food. They have to eat plants, or animals that have eaten other organisms.

The food that animals eat is called their DIET. If this contains all the necessary nutrients in the right amounts it is called a BALANCED DIET.

55% carbohydrates 15% fats	for release of energy
20% proteins	for repair and growth
10% water and salt	for health

In addition there must be ROUGHAGE (dietary fibre), to add bulk to the food, which helps to prevent constipation.

Energy content of food can be found by burning, e.g. a peanut and finding out the rise in temperature of a certain amount of water. From this can be calculated the energy content of different kinds of foods (see page 82).

Dietary requirements of different people:

1 Men need more energy than women per gram body mass.
2 Active occupations need most energy.
3 Young children need more energy per gram body mass than adults and they need more protein and calcium.
4 Pregnant and nursing mothers need extra energy and also more protein, calcium and vitamins for growth of the baby.

Healthy and unhealthy eating

(a) **Excesses,** e.g. excess food can cause obesity.

We need energy for all our body processes. Young children need a lot of energy to build up their bodies and grow. But adults only need enough energy for daily activities.

If the diet contains too much fat or carbohydrates the excess food will be stored as fat beneath the skin. If a person has a mass of more than 20 per cent above the average, then he or she is called OBESE. Obesity can increase a person's chance of having high blood pressure and a heart attack.

(b) **Deficiencies**

1 *Vitamins.* Vitamins are needed in very small amounts for normal functioning of the body. **Vitamin C** is found in citrus fruits such as lemons and oranges. A deficiency, i.e. an inadequate amount, can cause **scurvy** (bleeding gums). **Vitamin D** is found in fish oils. It is also made in the skin during sunlight. It is needed for healthy bone growth. A deficiency causes **rickets** (the bones are soft and can become bent).

2 *Roughage* (dietary fibre) is the indigestible cellulose plant cell walls found in vegetables, seeds and fruits and especially in bran and brown rice. It adds bulk to the food passing down the alimentary canal, aiding peristalsis. It is passed out as the undigested faeces. If there is insufficient roughage the person may become **constipated**.

3 *Protein.* Small children with insufficient protein needed for growth may develop **kwashiorkor**, with a swollen belly. If there is a great deficiency of food, the person could starve.

(c) **Food additives.** These are synthetic chemicals added to the food for a particular purpose. But they might cause harm.

1 *Preservatives*, e.g. sodium nitrite. This is added to processed meats such as bacon. It kills bacteria, and so helps to keep the food from going bad. But excessive nitrites can be dangerous to health, especially for small babies, who should not be given processed meats.

2 *Flavour enhancers,* e.g. monosodium glutamate. This is added to several foods used in Chinese cooking to improve the flavour. But a few people are sensitive to it, and have a reaction including headaches, dizziness and neck pains.

Summary

1 Nitrates are needed by plants to build proteins.
2 Plants and animals both need minerals, for example iron and calcium, for healthy growth.
3 Food stores of plants become the food supplies of animals.
4 A balanced diet contains carbohydrates, fats, proteins, water, mineral salts, vitamins and roughage in a certain proportion.
5 Excesses and deficiencies in the diet can be unhealthy.
6 Food additives can be useful, but may also cause problems.

Aims of the chapter

After studying this chapter you should be able to:

1 Describe the structure and care of the teeth.
2 Define digestion, absorption, assimilation and egestion.
3 List the enzyme actions in digestion of food groups.
4 Describe the parts and functions of the alimentary canal.
5 Describe the function of the villi of the small intestine.

Teeth structure, functions and care

V.S. of mammalian tooth

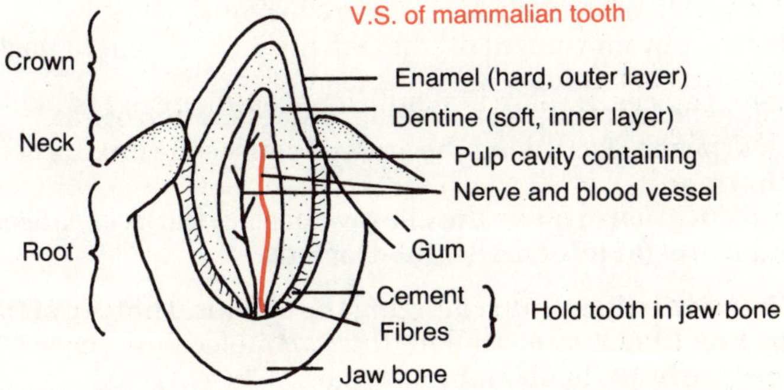

Teeth	*Structure and function*	
Incisors	Chisel-shaped crowns. Cut food.	Mechanical digestion: increases surface area.
Canines	Pointed crowns. Tear food.	
Premolars and molars	Crush and chew the food.	

Dental decay occurs when food (e.g. sugar) lodges between the teeth and is acted on by bacteria to produce acids. The acids eat into the enamel, and into the dentine. This is the cause of **tooth decay**. If the decay reaches the pulp cavity, pain is felt by the nerves (**toothache**). The decay is removed and a filling put in.

Care of the teeth:

1 Brush teeth to remove pieces of food.
2 Remove plaque (hardened food, bacteria and salts).
3 Crisp food (e.g. apples) help to clean the teeth.
4 Toothpaste is alkaline to neutralize bacterial acids.
5 Fluoride in water and toothpaste strengthens enamel.
6 Calcium and vitamin D in the diet help build strong teeth.

Definitions and examples

INGESTION: taking in of food into the alimentary canal through the mouth.

K

DIGESTION: the breakdown of large insoluble molecules into smaller molecules. It occurs in two stages:

1 **Mechanical digestion:** breakdown of food into smaller pieces of the *same* chemical nature. It increases the surface area of the food but does not change its nature.

Examples: cutting and chewing by the teeth produces a larger area for action of salivary amylase in the mouth.

Churning by movement of stomach muscles mixes stomach enzymes (in gastric juice) with the food.

Peristalsis (alternate contraction and relaxation of gut muscles) pushes food along the gut and causes the mixing of food and enzymes.

Emulsification of fat by the bile produced in the liver causes the breakdown of fat into smaller fat droplets.

2 **Chemical digestion:** changing the chemical nature of the food by breaking it into smaller, *different* molecules. These changes are brought about by enzymes.

Substrate	*Enzymes*	*Source*	*Products*
Proteins	Proteases	Stomach, pancreas, small intestine	Amino acids
Carbohydrates	Amylases	Mouth, pancreas, small intestine	Simple sugars
Fats	Lipases	Pancreas, small intestine.	Glycerol and fatty acids

ABSORPTION of food: uptake of soluble end-products of digestion into the bloodstream.

ASSIMILATION: use of and incorporation of the digested food into the body e.g. as an energy source (mainly sugars and fats), for storage (mainly fats) or for growth (mainly proteins).

EGESTION: expulsion of undigested remains of food (faeces) from the anus. (Water in the food is absorbed back into the blood in the colon. Faeces consist mainly of undigested cellulose (roughage), dead bacteria and worn-out cells.)

The alimentary canal and associated structures

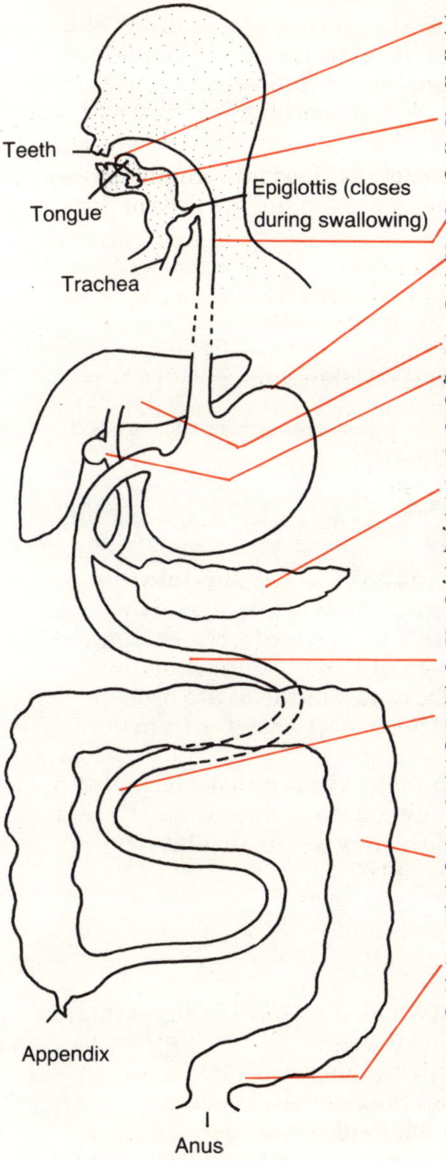

Mouth: food moistened and softened. Made into balls for swallowing.

Salivary glands: produce salivary amylase which begins the digestion of starch.

Oesophagus: tube to the stomach.

Stomach: produces acid to soften food; proteases to digest protein. Churning action breaks the food.

Liver: produces bile, stored in the *gall bladder*. Bile is alkaline. Bile duct takes bile to duodenum. Bile contains salts which break up (emulsify) fats.

Pancreas: produces pancreatic juice which is alkaline. *Pancreatic duct* takes pancreatic juice to duodenum. Juice contains enzymes acting on all food groups.

Small intestine consists of:
duodenum: receives bile and pancreatic juice for digestion.
ileum: long; place where most digestion and absorption occur. Intestinal juice contains enzymes acting on all food groups.

Large intestine consists of:
colon: absorbs water from the undigested food back into the blood.

rectum: stores dry, undigested food remains (faeces) until these are egested through the anus.

Absorption and Assimilation

The small intestine is very long (over 3 metres) and its inner wall is folded and covered with finger-like projections, **villi**. Thus food spends a long time in the small intestine to complete its digestion, and there is a large surface area for the absorption of digested food.

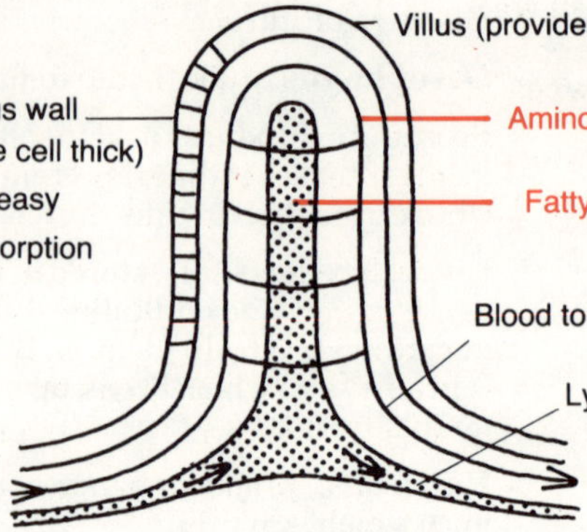

Simple sugars and **amino acids** go into the villus into the hepatic portal vein to the liver. Simple sugars are carried by blood to cells for respiration and release of energy. Excess sugars are changed to glycogen in the liver, or to fat. Amino acids go to the cells for growth and repair. Excess amino acids are broken down in the liver to make urea. Urea is later excreted from the kidneys in urine.

Fatty acids and **glycerol** go into the villus into the lacteal and the bloodstream. Used as an energy source in respiration. Excess fats are stored under the skin where they help to insulate the body against the cold. (See also page 107.)

Summary

1 Animals take in complex food which they have to digest (make soluble) before it can enter the bloodstream.
2 Digestion is mechanical (teeth) and chemical (enzymes).
3 Digestion occurs in the parts of the alimentary canal.
4 The liver produces bile which emulsifies fats.
5 The pancreas produces pancreatic juice containing enzymes.
6 Absorption occurs in the villi and food goes first to the liver.

For Questions **32–39**, each group of questions has five possible answers, (A), (B), (C), (D) or (E). For each question choose the best response. Each response may be used once, more than once or not at all.

Questions 32–35

(A) Cell membrane (B) Cell wall (C) Chloroplasts
(D) Chromosomes (E) Cytoplasm

Which parts of the cell listed above:
32 contain the hereditary information? ____
33 allow a plant cell to become turgid? ____
34 is made of cellulose? ____
35 can trap light energy? ____

Questions 36–39

(A) Carbon, hydrogen and oxygen (B) Nitrogen (C) Calcium
(D) Iron (E) Magnesium

Which of the elements listed above:
36 are found in all carbohydrates, fats and proteins? ____
37 is absorbed by plants in the form of nitrates? ____
38 is needed in the human diet to prevent anaemia? ____
39 helps to form hard bones and teeth? ____

40 A student set up the apparatus shown below to investigate the effect of amylase on the digestion of starch.

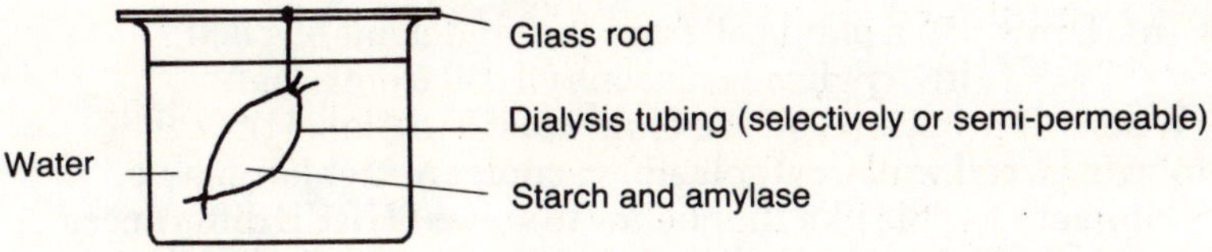

(a) What apparatus should the student set up as a control?
(b) (i) What test should the student use to find out if the enzyme has worked?
(ii) Which liquid should the student be testing?
(iii) What reaction has taken place?
(c) Give TWO ways in which the apparatus is:
(i) similar to the small intestine.
(ii) different from the small intestine.

41 (a) What is the active ingredient of biological washing powders?
(b) At which of the following temperatures is the biological washing powder likely to work the fastest?
(A) 0°C (B) 10°C (C) 20°C (D) 30°C (E) 50°C
(c) Give a reason for your answer to part (b) above.

42 (a) List FIVE characteristics of enzymes.
(b) Choose one of the characteristics and describe briefly how you would demonstrate it in the lab.

43 Describe how would you test:
(a) for starch.
(b) for oxygen.
(c) for carbon dioxide.
(d) for glucose.

44 Which of the following substances is NOT needed by plants for photosynthesis?
(A) Light (B) Chlorophyll (C) Oxygen
(D) Carbon dioxide (E) Enzymes

45 On a bright summer's day the increase in which of the following factors is likely to increase photosynthesis by the greatest amount?
(A) Oxygen (B) Carbon dioxide (C) Water
(D) Light (E) Heat

46 (a) Draw (i) a plant cell from the epidermis of a leaf.
(ii) a palisade mesophyll cell from a leaf.
(b) Where appropriate, on the diagrams in (a), label the cell membranes, cell walls, cytoplasm, vacuoles and chloroplasts.
(c) Complete a table like that below to show THREE differences between a typical animal cell and a palisade mesophyll cell.

	Typical animal cell	*Palisade mesophyll cell*
(i)		
(ii)		
(iii)		

47 A certain breakfast cereal contains 28.6 per cent fibre by mass.
(a) How many grams of fibre would there be in 100 grams of the cereal?
(b) How many grams would there be in an average serving of 40 grams of the cereal? Show your working.
(c) Why is fibre important in the diet?

48 The table below shows the composition of five foods in our diet.

	Composition per 100 g edible part					
	Proteins	*Fat*	*Carbohydrates*	*Calcium*	*Iron*	*Vitamin C*
(A)	2 g	—	19 g	11 mg	0.7 mg	15 mg
(B)	18 g	5 g	—	10 mg	12 mg	—
(C)	1.5 g	—	3 g	40 mg	0.5 mg	30 mg
(D)	0.6 g	82 g	0.4 g	16 mg	—	—
(E)	0.2 g	—	96 g	30 mg	2 mg	—

(a) (i) Which of the foods is most likely to be cabbage? ____
(ii) Give two reasons for your choice.
(b) (i) Which of the foods would be MOST useful for a pregnant woman? ____
(ii) Give three reasons for your choice.
(c) The energy per gram (kJ/g) provided by each of the food groups is shown below:

Carbohydrates	17 kJ/g
Proteins	18 kJ/g
Fats	39 kJ/g

(i) Which of the foods, (A), (B), (C), (D) or (E) in the table above would provide the most energy per 100 grams?
(ii) Explain your answer.
(d) (i) What disease is caused by a deficiency of vitamin C?
(ii) Give one symptom of the disease.

49 Give a function related to digestion of each of the following:
(a) Incisors (b) Molars (c) Saliva (d) Oesophagus
(e) Stomach (f) Liver (g) Pancreas
(h) Small intestine (i) Colon (j) Rectum

50 Define and give examples of:
(a) Mechanical digestion (b) Chemical digestion

51 The ingredients of two different brands of raspberry yoghurt are listed below.

Brand A		*Brand B*	
per 100 g		per 100 g	
Energy	172 kJ	Energy	452 kJ
Protein	4.8 g	Protein	4.1 g
Fat	0.2 g	Fat	2.5 g
Carbohydrate	5.5 g	Carbohydrate	18.4 g
Artificial sweetener			

(a) (i) Which of the two brands would a slimmer choose to eat?
(ii) Give three reasons for your choice.
(b) (i) What is yoghurt made from?
(ii) Neither of the yoghurts contains preservatives. Where should they be stored to stop them going bad?
(iii) Could you keep them from ever going bad? Explain your answer.
(c) A carton of yoghurt contains 125 g. For brand A, how much of the following substances would there be in a carton?
(i) Protein (ii) Fat Show your working.

52 (a) Digestive enzymes are divided into three main groups. Complete a table like that below to give the substrates and end-products for each group.

Enzyme groups	*Substrate (substance digested)*	*End-product (end result)*
Proteases		
Lipases		
Amylases		

(b) (i) A certain enzyme is most active in acid conditions. Which of the following graphs would show its activity?
(ii) Explain your answer.

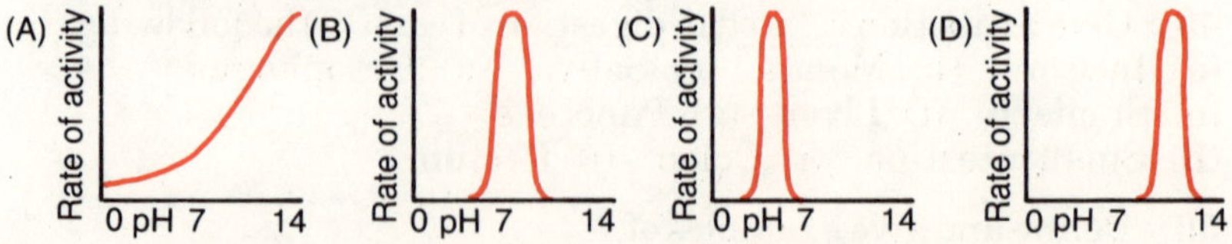

Now turn to page 145 to check your answers 32–52.

15 Transport systems

Aims of the chapter

After studying this chapter you should be able to:

K

1 Explain the need for transport systems.
2 Describe transport systems in plants, including the distribution of xylem and phloem in a stem and in a root.
3 Describe the composition and functions of blood.
4 Describe heart structure and the circulatory system.
5 Compare the structure and functions of arteries and veins.
6 Discuss health aspects of the circulatory system.

The need for transport systems

A unicellular organism such as *Amoeba* has a large surface area in comparison to its volume. The gases it needs to take in and to get rid of, can easily diffuse through its surface. (For an explanation of diffusion look at page 53.)

A larger organism has less surface area in comparison with its volume. It will probably also be more active and cannot depend upon diffusion, which is very slow, for movement of gases and water. A system for the mass movement of large amounts of material (MASS FLOW) within vessels is needed in a large organism, such as a flowering plant and a mammal.

K

Water intake in plants

K

The roots of a plant have very many projections, the ROOT HAIRS. The root hairs increase the surface area in contact with the soil through which water (and mineral salts) are taken in. Water enters the root hair by OSMOSIS as the cell sap is more concentrated than the soil water. Water passes from cell to cell across the root by osmosis, and then reaches the xylem vessels.

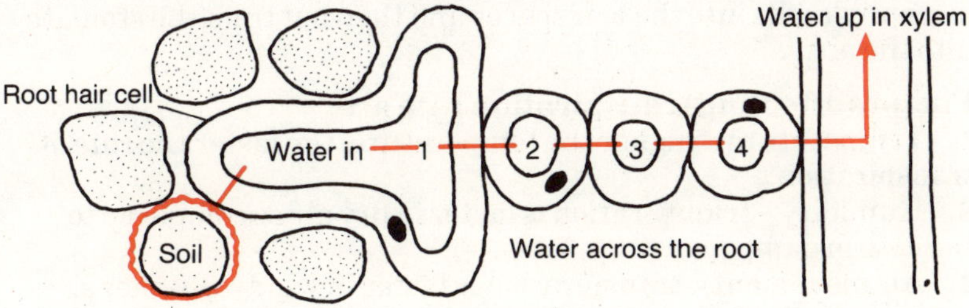

Water up the plant

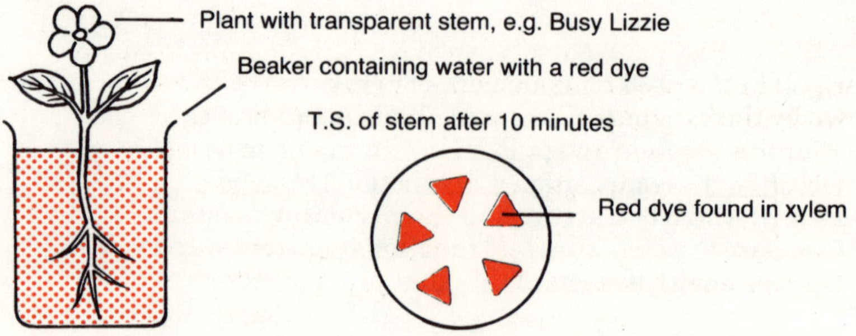

Water is transported in the xylem vessels. It is pushed up from below (as more water enters the roots), and is pulled up from above (by transpiration).

Water out of the plant (Transpiration)

K

TRANSPIRATION is the loss of water vapour (mainly from the lower surface of leaves). This loss of water pulls more water up through the xylem. Note: The plant WILTS (becomes droopy) if water loss is greater than water intake.

Water loss by transpiration is shown by:

1 Putting a plant inside a closed plastic bag. Drops of water will form on the inside of the bag.

2 Putting dry (blue) cobalt chloride paper onto leaves. When damp this becomes pink.

3 Attaching a plant to a POTOMETER. The uptake of water by the plant is recorded as the movement of water along a scale. Uptake should be a good measure of water loss by transpiration.

Transpiration occurs as water vapour diffuses from the spongy mesophyll cells into the leaf spaces and then out from the stomata into the air.

Factors affecting transpiration rate are:

1 Temperature – higher the temperature, the faster the rate of transpiration.

2 Humidity – transpiration is faster in dry air as compared to the rate in damp air.

3 Air movement – transpiration is faster in windy weather as compared with the rate in still air.

Movement of food in the plant (Translocation)

Dissolved food (mostly sugars and amino acids) is transported in the phloem. This transportation or TRANSLOCATION needs energy. The downward movement of food from the leaves is shown by the 'ringing experiment' illustrated below.

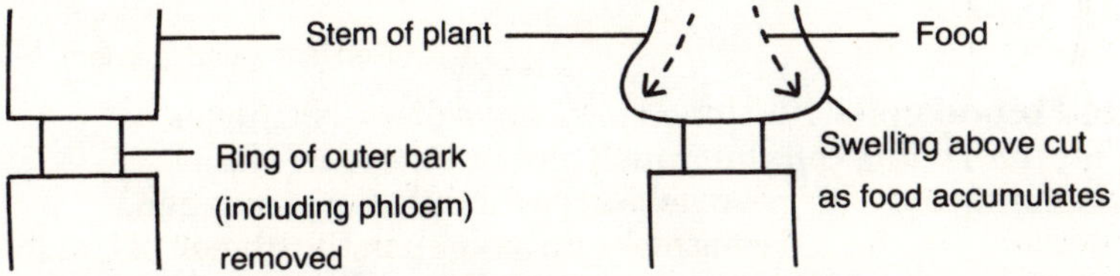

Distribution of xylem and phloem in a dicotyledon

T.S. root

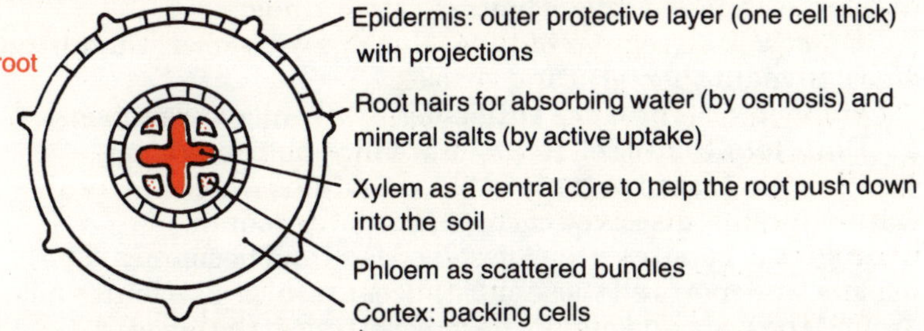

T.S. stem

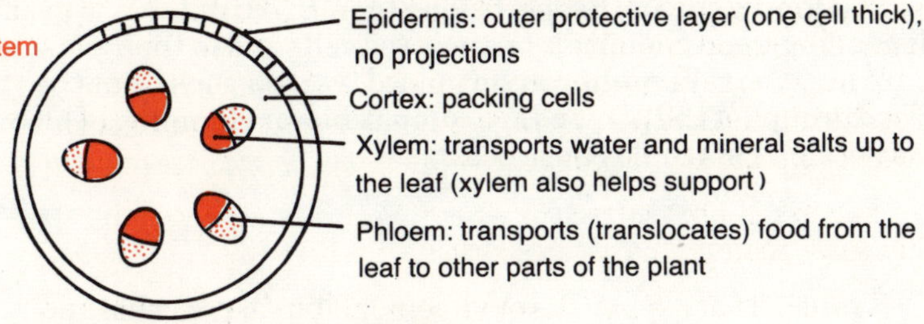

Summary: Transport systems in plants

1 Water is taken in by root hairs by osmosis.
2 Mineral salts are taken in by active absorption.
3 Water and mineral salts travel up the xylem.
4 Water is lost from the plant by transpiration.
5 Food substances are translocated in the phloem.

Blood composition and functions

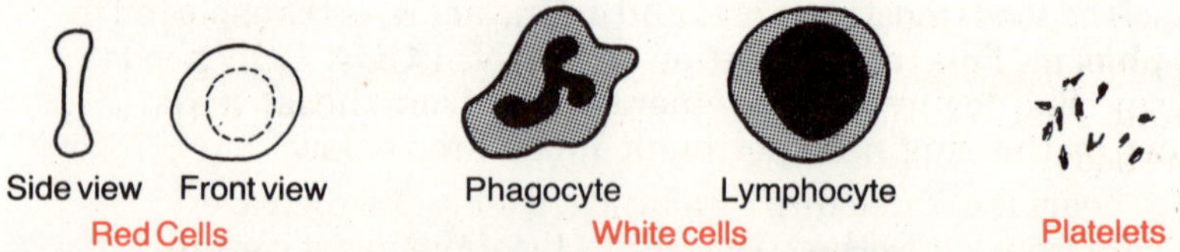

1 Red blood cells. Flattened biconcave discs. No nuclei. Smaller than white cells but much more numerous. Have haemoglobin and a large surface area for diffusion of oxygen. Haemoglobin in the red cells picks up oxygen in the alveoli of the lungs and transports it to the body cells. The red cells circulate in the blood until they return to the lungs.
2 White blood cells. Spherical. With nuclei. Larger, but less numerous than red cells. They are of two kinds:
Phagocytes – irregular outline. Nucleus with lobes. Engulf and digest invading bacteria and viruses.
Lymphocytes – produce antibodies to kill micro-organisms.
3 Platelets. Small cell fragments which help in clotting.
4 Plasma. Liquid part of the blood contains water, mineral salts, vitamins, dissolved carbon dioxide (as sodium hydrogencarbonate), urea, dissolved food and hormones. The plasma transports all these substances and also distributes heat around the body and plays a part in clotting of the blood.

Note. **Blood groups.** People belong to A, B, AB or O blood groups depending upon chemicals on their red cells and in their plasma. Care must be taken when giving blood transfusions, that the blood groups of the persons are compatible, or clumping of blood may occur. This could cause death.

Summary: Functions of the blood

1 Transport of oxygen (as oxyhaemoglobin in red cells) and carbon dioxide (dissolved in plasma); and food, urea, salts, vitamins and hormones in the plasma.
2 Defence: phagocytes ingest bacteria and viruses, lymphocytes produce antibodies, platelets and plasma cause clotting of the blood to seal cut blood vessels.
3 Temperature regulation: distribution of heat around the body, and gain or loss of heat by blood vessels in the skin.

The circulatory system

The circulatory system consists of the vessels (arteries, veins and capillaries) and a pump (the heart) which pushes the blood around. There is a double circulation – to the lungs and to the rest of the body.

The heart is divided into right and left halves which are separate from each other. The right side of the heart contains DE-OXYGENATED blood (much carbon dioxide, little oxygen). The left side of the heart contains OXYGENATED blood (much oxygen, little carbon dioxide).

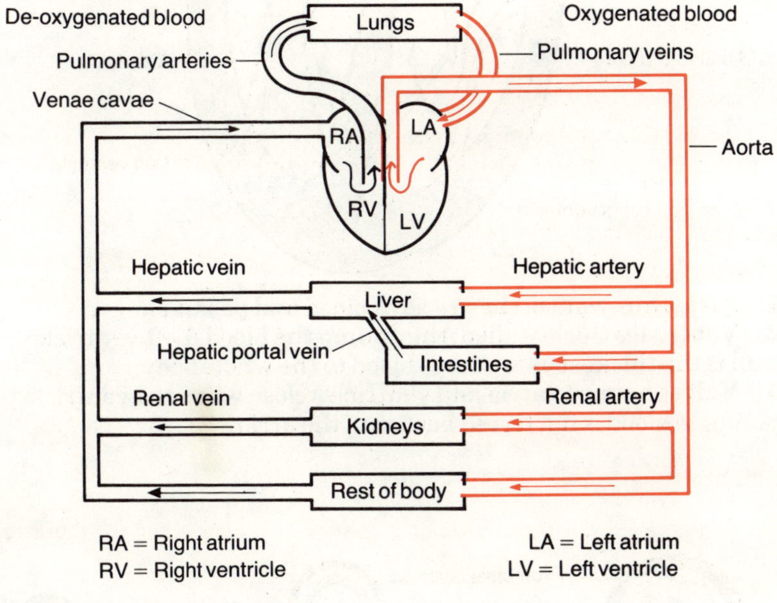

Summary: Blood circulation

1 Arteries take blood away from the heart.
2 Veins take blood towards the heart.
3 Capillaries take blood to and from body cells.
4 De-oxygenated blood is found in most veins; in the right side of the heart and in the pulmonary *arteries*.
5 Oxygenated blood is found in most arteries; in the left side of the heart and in the pulmonary *veins*.

Structure of the heart (ventral section)

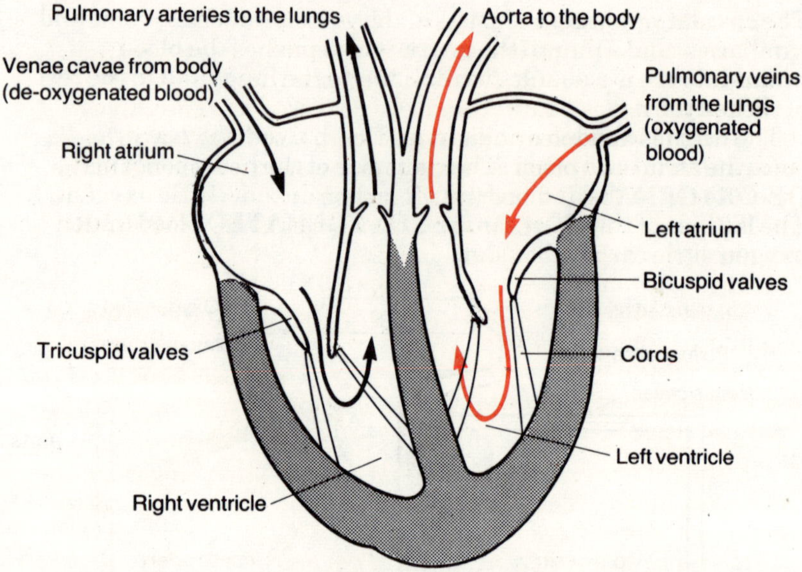

1 Atria thin-walled: they receive blood and pass it on.
2 Ventricles thick-walled: they pump the blood. Left ventricle wall is the thickest as it pumps blood to the whole body.
3 Valves between atria and ventricles close when the ventricles pump, so blood is not forced back into the atria.

Structure of arteries, veins and capillaries

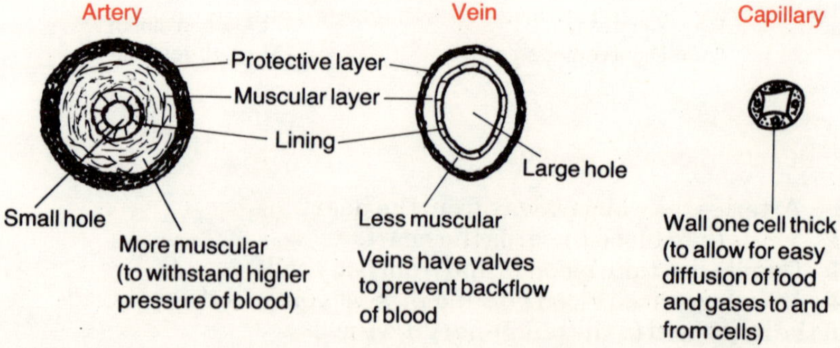

Blood circulation and health

1 Healthy circulation

(a) *Heart beat*. When the heart relaxes blood is drawn in from the veins into the atria, and then into the ventricles.

When the heart 'beats', the ventricles contract. The valves between the ventricles and atria close (so blood will not go back into the atria) and blood is forced into the arteries and out of the heart to the lungs and the rest of the body.

(b) *Pulse*. The 'push' of the ventricles is felt as the PULSE, the surges of blood in the arteries. In an adult the heart beats about 70 times per minute, so the pulse rate is 70.

(c) *Valves*. Blood is pushed strongly along the arteries to the capillaries which supply the tissues. When blood is collected up into the veins it is at a very low pressure. Valves in the veins stop backflow of blood. Damaged valves cause varicose veins where blood accumulates in swellings, especially in the legs.

2 Effect of exercise. When exercising the breathing rate increases, and the heart beats more quickly – and so the pulse is increased. More oxygen and food are sent to the muscles. After exercise the pulse quickly returns to normal.

3 High blood pressure. If the arteries become narrowed as a result of smoking, too much saturated fat in the diet, or too much stress, the heart has to work harder to push the blood around the body. This causes high blood pressure.

4 Coronary arteries. The muscles of the heart are supplied with food and oxygen by the CORONARY ARTERIES. If these become narrowed (as described above), the supply is interrupted. If they are very narrow and prevent oxygen getting to the heart muscles it could cause angina or a heart attack.

Summary

1 Xylem (dead vessels) transport water and mineral salts.
2 Phloem (living cells) transport (or translocate) food.
3 Blood contains red and white cells, platelets and plasma.
4 Blood is collected from the lungs and body in veins, and distributed to the lungs and body in arteries.
5 The heart contains atria (which receive blood) and ventricles (which pump blood to the lungs and body).

16 How respiration releases energy

Aims of the chapter

After studying this chapter you should be able to:

1. Describe and explain the importance of respiration.
2. Distinguish between aerobic and anaerobic respiration.
3. Give examples of gaseous exchange.
4. Describe the human respiratory system.
5. Discuss health aspects of the respiratory system.

K

The importance of respiration

K

RESPIRATION is the energy-releasing process in all living cells at all times. The energy that was trapped in the food during photosynthesis is released.

Energy is used for:

1. All the processes of 'staying alive'.
2. Movements within the protoplasm, and muscle contraction.
3. Active transport, e.g. uptake of mineral salts.
4. Keeping a constant body temperature in some animals.
5. Building up (synthesis), e.g. of proteins for growth.
6. Cell division and reproduction.

Energy content of food can be found e.g. by burning a peanut and finding the increase in temperature of a certain mass of water, as shown below.

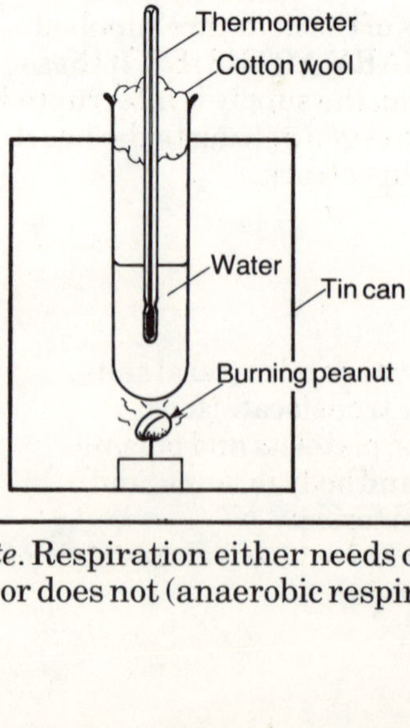

Note. Respiration either needs oxygen (aerobic respiration, page 83) or does not (anaerobic respiration, page 84).

Aerobic respiration

AEROBIC RESPIRATION requires oxygen. It releases much more energy than anaerobic respiration. The food substance is completely burned (combusted) to release all its energy.

Glucose + Oxygen → Energy + Carbon dioxide + Water

1 **Oxygen is used up.** If respiration is allowed to occur in a closed container, then the air at the end of the experiment will not allow a candle to burn in it: no oxygen is left.

2 **Energy is released.** Germinating seeds are put in a Thermos flask, and dead seeds in another one. All the seeds are first washed with disinfectant to kill any bacteria which would also be respiring. Thermometers in the flasks show a temperature rise in the living, respiring seeds but not in the dead seeds. So respiration releases energy.

3 **Carbon dioxide and water are released.** A piece of biscuit is held by tongs in a bunsen flame until it burns. The gases produced are collected in a gas jar. When dry (blue) cobalt chloride paper is added it goes pink, showing water is present. When limewater is added it goes cloudy, showing carbon dioxide has also been produced in respiration.

4 **Inspired and expired air** are compared as shown below.

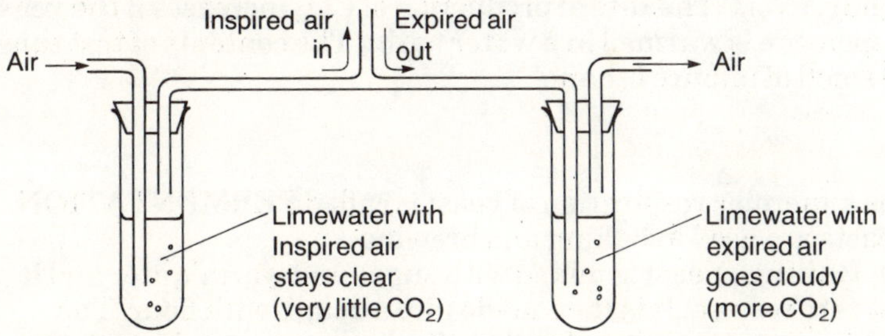

Other experiments and accurate measurements show:

Inspired air contains 80% nitrogen, 20 % oxygen, 0.03% carbon dioxide, little water vapour and has a temperature less than 37°C.

Expired air contains 80% nitrogen, 16% oxygen, 4% carbon dioxide, much water vapour and has a temperature of 37°C.

Anaerobic respiration

K

ANAEROBIC RESPIRATION takes places without oxygen. It releases less energy than aerobic respiration.

(a) **In animal muscles** during exercise, **lactic acid** is produced. This is the reason for muscle fatigue. After exercise there is an oxygen debt, and more oxygen is needed to convert the lactic acid to harmless substances.

Glucose → Lactic acid + Energy

(b) In **germinating seeds** and **yeast**, alcohol **(ethanol)** and carbon dioxide are produced.

Glucose → Ethanol + Carbon dioxide + Energy

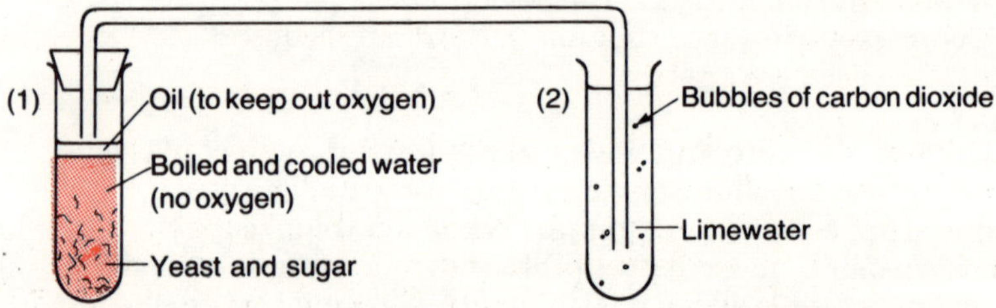

Results Test tube (1) becomes warm as energy is released. Bubbles of carbon dioxide make the limewater in test tube (2) go cloudy. (*Note*. The rate of production of CO_2 increases if the yeast suspension is warmed in a water bath.) The contents of test tube (1) smell of impure ethanol.

Uses of anaerobic respiration

K

The anaerobic respiration of yeast is called FERMENTATION. Yeasts are used in baking and brewing.
(a) **Baking.** Yeast is mixed with sugar and warm water and it starts to respire. It is then made into a dough with flour. The carbon dioxide produced makes the dough rise. Cooking the bread kills the yeast, and the gas is trapped inside.

(b) **Brewing** for making wines and beers makes use of the fact that fermentation produces ethanol. A concentration of over 14% ethanol kills the yeast. Spirits such as rum, whisky and gin are distilled to increase their ethanol concentration.

Gaseous exchange

(a) **Characteristics of gas exchange surfaces**

1 Large surface area
2 Thin
3 Moist
4 Permeable

} Allow gases to move easily through the surface along their diffusion gradients

(b) **Gaseous exchange in flowering plants**

Large surface area of spongy mesophyll cells in the leaf (see page 61). The cell walls are thin, moist and permeable and allow water vapour to diffuse from the cell walls into the spaces close to the stomata, and then out into the air. Oxygen and carbon dioxide also diffuse in or out along their diffusion gradients depending upon the balance between respiration and photosynthesis as shown below.

Daytime or light
Respiration: O_2 in, CO_2 out
Photosynthesis: O_2 out, CO_2 in

Night-time or dark
Respiration: O_2 in, CO_2 out

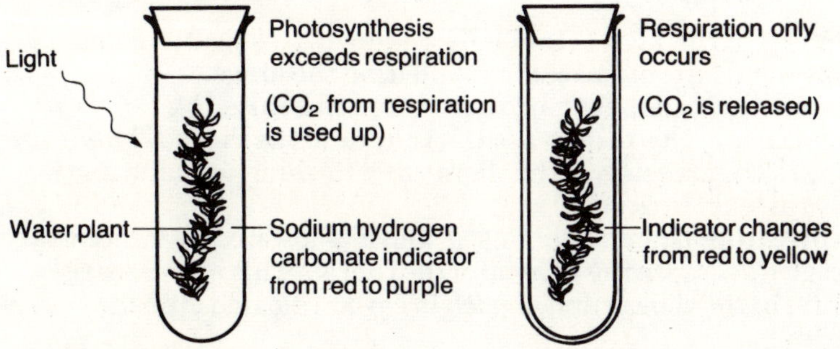

(c) **Gaseous exchange in a mammal**

The lungs (see page 86) consist of branched tubes which end in millions of sacs called alveoli. There is therefore a very large surface area for gaseous exchange. The alveoli walls are permeable and only one cell thick, and there is a layer of moisture in which the gases can dissolve. De-oxygenated blood from the heart in the pulmonary arteries is brought close to the alveolus. **CO_2 diffuses into the alveolus, from the blood. O_2 diffuses into the blood from the alveolus.** The oxygenated blood is then returned to the heart in the pulmonary veins (see pages 79 and 86).

Human respiratory system

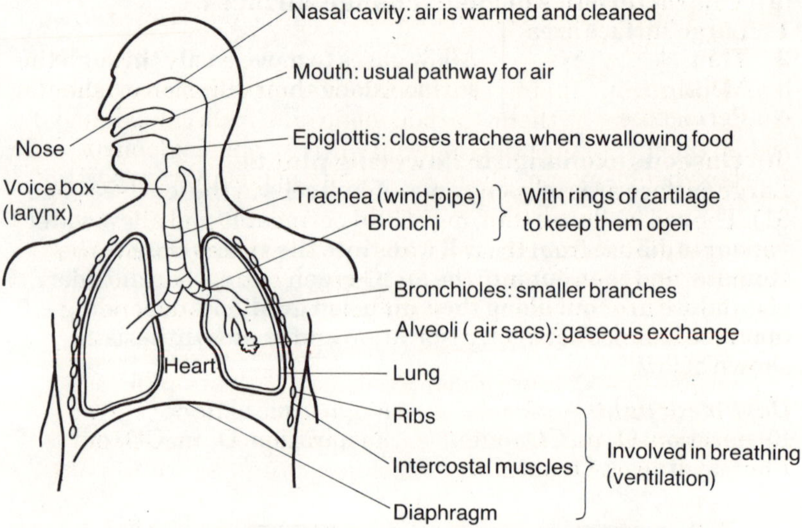

K

BREATHING or VENTILATION: taking in (Inspiration) and passing out (Expiration) of air to and from the lungs.

Inspiration: Diaphragm lowered, and ribs raised by intercostal muscles to increase the volume of the thorax. This decreases the pressure in the thorax and the lungs expand and draw in air rich in oxygen.

Expiration: Diaphragm raised, and ribs lowered by intercostal muscles to decrease the volume of the thorax. This increases the pressure in the thorax and air rich in carbon dioxide is pushed out of the lungs.

Gaseous exchange in the alveolus

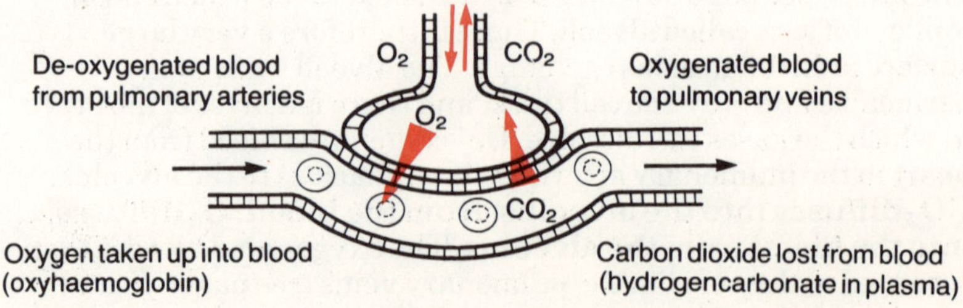

Respiratory system and health

(a) **Exercise.** During exercise we breathe more quickly and more deeply to draw more oxygen into the lungs. The alveoli open up and the capacity of the lungs increases.

(b) **Pollution.** Sulphur dioxide (in acid rain) can irritate the lungs and damage them. Carbon monoxide (from car exhausts) when taken into the lungs joins up with the haemoglobin (to form carboxyhaemoglobin) and stops the haemoglobin transporting oxygen. In large amounts it can kill.

(c) **Disease.** Bacteria and viruses are transferred in the air we breathe. Some of these affect the lungs and cause bronchitis. This is made worse by smoking and air pollution.

K

(d) **Smoking.** Cigarette smoke contains tar, nicotine and carbon monoxide. Smokers are much more likely to develop LUNG CANCER than non-smokers. The greater the number of cigarettes smoked, the greater is the risk. (*Note*. The risk of getting lung cancer gradually decreases when cigarette smoking is stopped.)

There is a combined effect of cigarette smoking and air pollution. Smokers in polluted areas have the highest rate of lung cancer, and non-smokers in low pollution areas the lowest.

Smoking more than 20 cigarettes a day doubles the chance of death from heart attacks in men 50–70 years old.

Long-standing bronchitis can also damage the alveoli so that gaseous exchange is less efficient and the person becomes 'breathless'.

Mothers who smoke a lot of cigarettes have babies who are smaller than average and at more risk from infection.

Summary

1 Respiration releases energy for life processes.
2 Aerobic respiration needs oxygen (most organisms)
Glucose + Oxygen → Energy + Carbon dioxide + Water
3 Anaerobic respiration: no oxygen, less energy released
Muscles: Glucose → Lactic acid + Energy
Yeast: Glucose → Ethanol + Carbon dioxide + Energy
4 Gaseous exchange occurs in all organisms.
5 Pollutants and smoking can damage the respiratory system.

17 Getting rid of wastes

Aims of the chapter

After studying this chapter you should be able to:

1 Define excretion and distinguish it from egestion.
2 List excretory products in plants and animals.
K 3 Describe the human excretory system.
4 Discuss health aspects of the excretory system.

Excretion and egestion

K EXCRETION is the removal of waste products of metabolism. (*Note*. Metabolism is the chemical reactions in the body.)

For example: Respiration is a chemical reaction which produces carbon dioxide, water and heat which are excreted. Excess amino acids are broken down in the liver to form UREA. This is taken to the kidneys where it is excreted.

K EGESTION is the removal of undigested remains of food, the FAECES. They have never been in the body cells.

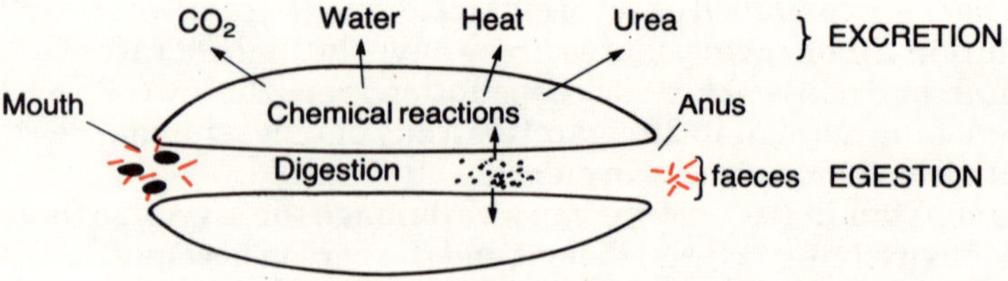

Excretory organs	Excrete
Lungs	CO_2, H_2O vapour, heat
Skin	H_2O vapour, heat, salts
Kidneys	Urea, excess H_2O and salts, some heat

Excretion in plants

In the *daytime*: Photosynthesis exceeds respiration (see page 85); the CO_2 is used up, but excess O_2 is excreted.

In the *nighttime*: Only respiration occurs; CO_2 and water vapour are excreted.

Plants are less active than animals and they excrete less wastes. Plants only make proteins as they are needed, so they do not have nitrogen-containing excretory products.

Human excretory system

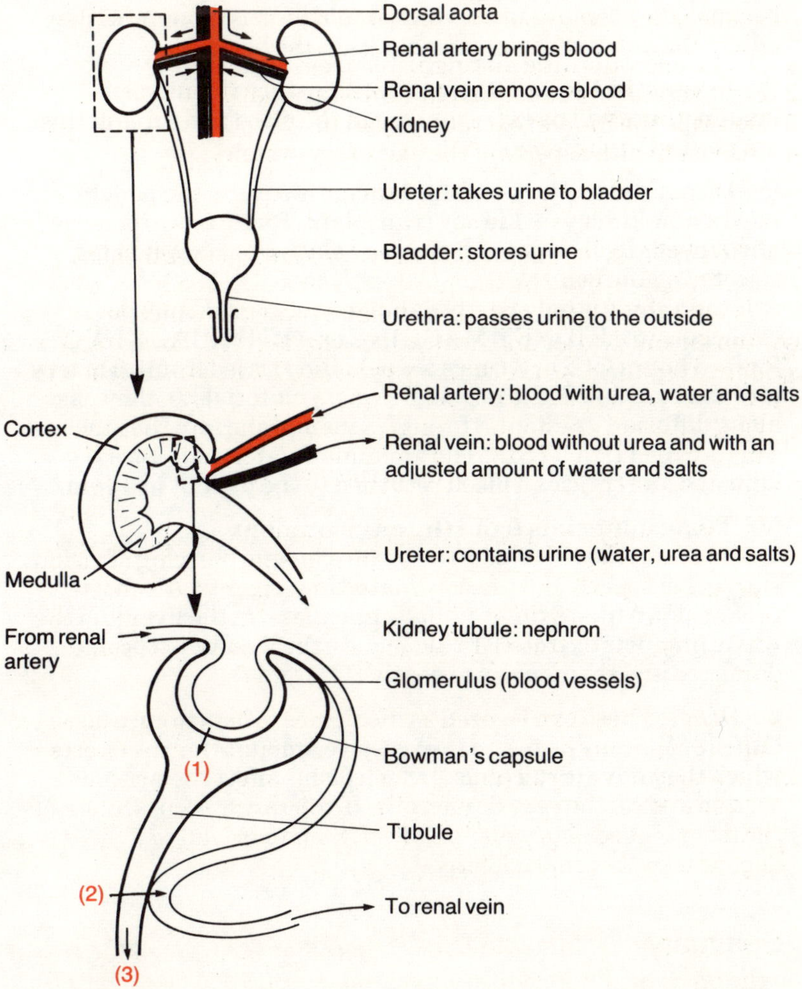

1 Filtration: Water, glucose, urea and salts pass from the blood into the tubule.
2 Selective reabsorption: Glucose; and some water and salts are taken back into the blood (osmo-regulation).
3 Urine (urea, water and salts) is passed to the bladder.

Excretory system and health

(a) **Normal functioning.** The kidneys are essential for life because they remove urea which would be poisonous, and they adjust the amount of water and salts in the blood.

(b) **'Sugar' in the urine.** If a person produces insufficient insulin, there will be extra glucose in the blood and some of this may be lost in the urine.

(c) **Damaged kidneys.** If the damage is severe, the person may need a new kidney – a kidney transplant. But this is expensive, and it needs to be donated by a near relative to avoid it being rejected by the body.

K If the kidney is only partly damaged the person may be connected to a KIDNEY MACHINE or ARTIFICIAL KIDNEY for 6–8 hours for 2 or 3 times a week. Blood is led from an artery into the machine. Here it loses its waste materials as they pass along diffusion gradients through a membrane into the machine. K This is called DIALYSIS. The amount of water and salts is adjusted and the clean blood returned to the patient in a vein.

(d) **Toxic materials from the environment**

1 *Alcohol.* Alcohol affects the brain, causing blurring of vision, slurring of speech and uncoordinated movements. Alcohol is broken down (de-toxified or made harmless) in the liver and the waste products excreted. In alcoholics, the liver may become damaged and this can cause death.

2 *Drugs.* These can be used as medicines to help to cure disease. Other drugs may be taken in excessive amounts for the effects which they have (drug abuse). Such drugs affect the nervous system and can be very dangerous. If 'tolerance' occurs, more of the drug is needed to cause an effect. Some can lead to dependency (become addictive).

Summary

1 Excretion is removal of waste products of metabolism.
2 Plants excrete O_2 in the daytime and CO_2 at night.
3 The lungs, skin and kidneys excrete CO_2, H_2O, heat, salts and urea. Of these the kidneys are the most important.
4 A kidney machine can take over the functions of a kidney.
5 The body gets rid of toxic materials from the environment.

53 The diagram below shows the blood circulation in a mammal.

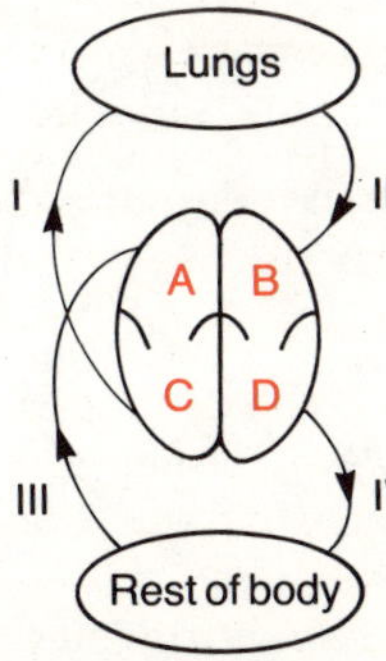

(a) (i) Which of the chambers has the most muscular walls? What is the importance of this?
(ii) Which chambers of the heart contain de-oxygenated blood?

(b) (i) Name the main blood vessels associated with the heart.
I II III IV
(ii) Which vessels contain oxygenated blood?
(iii) Which TWO of the following statements are correct?
1 Arteries always carry oxygenated blood.
2 Veins always carry de-oxygenated blood.
3 Arteries always carry blood away from the heart.
4 Veins always carry blood towards the heart.

(c) (i) From where does the heart muscle itself receive blood?
(ii) If these vessels are narrowed or blocked, what may be two of the results?

Questions **54–57** have five possible answers: (A), (B), (C), (D) or (E). For each question choose the best response. A response may be used once, more than once or not at all.

(A) Red blood cells (B) Lymphocytes (C) Phagocytes
(D) Platelets (E) Plasma

Which part of the blood:
54 Carries oxygen as oxyhaemoglobin? ____
55 Carries carbon dioxide as bicarbonate (hydrogencarbonate) ions? ____
56 Produces antibodies against certain bacteria? ____
57 Carries hormones around the body? ____

58 The diagram below shows a T.S. of a dicotyledonous stem.

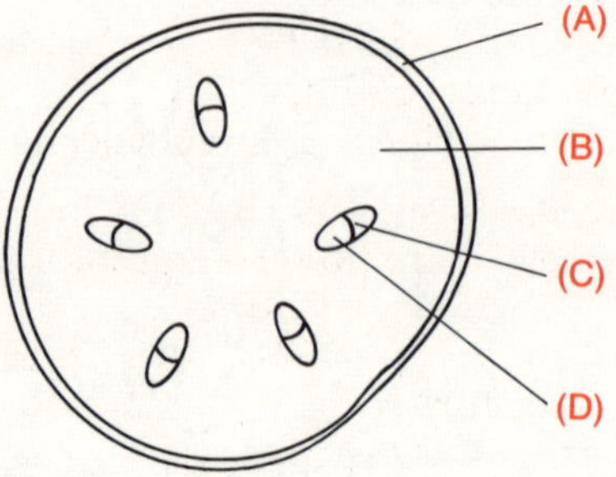

(a) In which one of the tissues, (A). (B), (C) or (D) do water and mineral salts rise up in the plant?
(b) In which of the tissues would food be moved down?
(c) What are the names of the processes by which:
 (i) Food is made in the plant?
 (ii) Food is moved through a plant?
 (iii) Water is taken in by a plant?
 (iv) Water is lost by a plant?
(d) (i) List THREE conditions which would increase the rate of water loss from a plant.
 (ii) Explain the effect of one of these conditions.
(e) What happens to the plant if water loss is greater than water uptake?

Questions **59–64** have six possible answers, (A), (B), (C), (D), (E) or (F). For each question choose the best response. A response may be used once, more than once or not at all.

(A) Excretion (B) Defaecation (C) Urethra
(D) Ureter (E) Liver (F) Kidney

59 The removal from the body of undigested food is known as ____

60 The removal from the body of metabolic wastes is known as ____

61 Urea is made in the ____

62 Urea is taken from the kidneys in the ____

63 A dialysis machine supplements the functioning of the ____

64 Toxic materials from the environment are detoxified in the ____

65 Given below are the equations for three energy-releasing processes.

(A) Glucose + Oxygen → Carbon dioxide + Water + Energy
(B) Glucose → Lactic acid + Energy
(C) Glucose → Carbon dioxide + Ethanol (alcohol) + Energy

(a) Which of the processes releases the MOST energy? ____
(b) Which of the processes are called anaerobic respiration? ____ and ____
(c) (i) Mention one place where process (B) could occur.
(ii) Why does it occur here?
(iii) What is the result of the process?
(d) (i) Mention one place where process (C) could occur.
(ii) Why is this reaction important?

66 (a) List THREE characteristics of a respiratory surface common to both plants and animals.
(b) Describe gaseous exchange in a flowering plant, saying how these characteristics are important.
(c) (i) List TWO further characteristics helping gaseous exchange in a mammal.
(ii) Why are these extra characteristics necessary?

67 The diagram below shows the apparatus used to compare the carbon dioxide in inhaled and exhaled air.

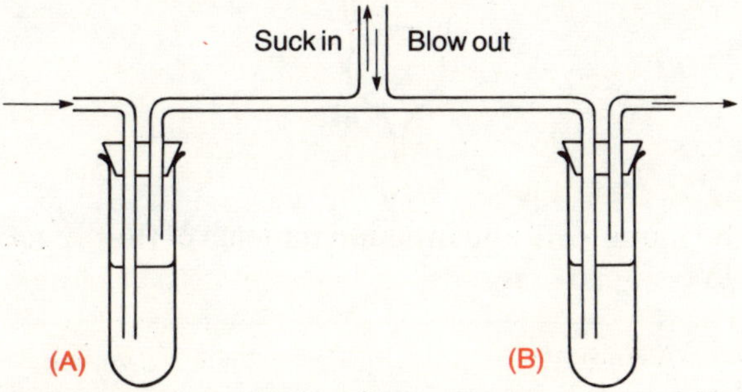

(a) (i) Is inhaled or exhaled air passed through test tube (A)?
(ii) What is the purpose of test tube (A)?
(b) (i) What should be the liquid in test tubes (A) and (B)?
(ii) What happens to the liquid in test tube (B)?
(iii) Why does this happen in test tube (B)?

68 The diagram below shows an alveolus and its blood supply.

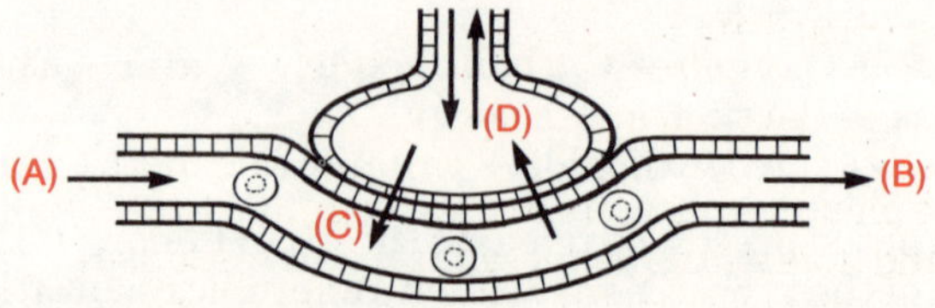

(a) Correctly fill in the spaces in the following sentences.
Blood is brought from the (i) ____ to (A).
(ii) ____ gas is taken up along the diffusion gradient (C).
(iii) ____ gas is lost along the diffusion gradient (D).
(iv) Blood leaves at (B) to go to the ____.

(b) List THREE likely effects of heavy cigarette smoking.

69 The diagram below shows part of the excretory system of a human male, together with the blood supply.

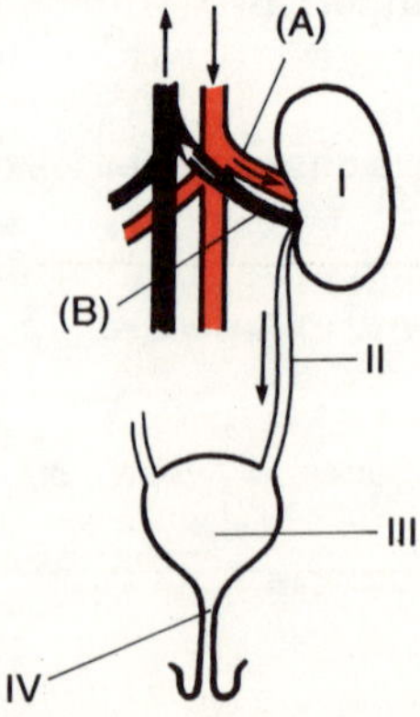

(a) Give the name, and one function for each of the structures I, II, III and IV.

Name	*Function*

(b) List THREE ways in which the blood in vessel (A) will differ from the blood in vessel (B).

Now turn to page 148 to check your answers 53–69.

18 Sensing and responding to changes

Aims of the chapter

After studying this chapter you should be able to:

1 Describe phototropism, its cause and importance.
2 Define and give examples of stimuli and responses.
3 Distinguish between voluntary and reflex actions.
4 Describe an example of a reflex arc and its importance.
5 Describe the structure and functioning of the eye.
6 Give a simple account of the importance of the ear.

Phototropism in plants

PHOTOTROPISM is a growth movement towards light coming from one direction. The stem grows (or bends) towards the light. *Importance*. This **response** to the **stimulus** of light makes sure that the plant gets light for photosynthesis.

Young seedlings are used. They are treated as shown below.

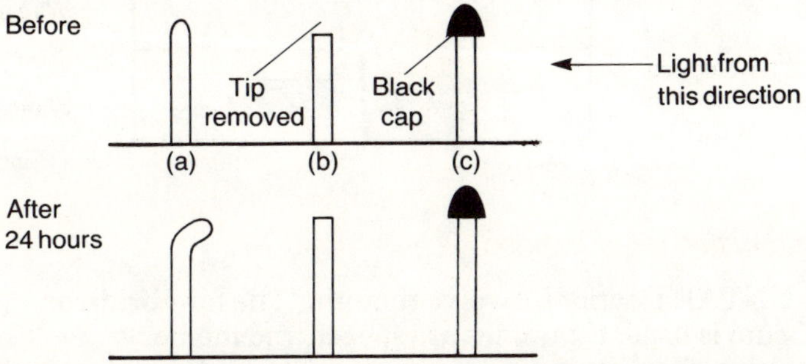

(a) Shoot bends towards the light. A growth hormone, AUXIN, is produced in the tip. This accumulates on the side *away* from the light (the left side in this case) and makes it grow more and so the shoot bends towards the light.
(b) Shoot grows straight upwards. The tip which would have produced auxin has been removed.
(c) Shoot grows straight upwards. The tip is covered, so light cannot affect it to cause auxin to be made.

Note. Plants are also affected by gravity (geotropism). Roots grow towards gravity (and so they grow down into the soil to anchor the plant). Shoots grow away from gravity (and so they grow up into the light for photosynthesis).

Stimuli and responses

K

STIMULI are changes in external or internal conditions that affect an organism. For example: light from one direction is the stimulus which causes phototropism.

K

RESPONSE is the action that occurs as a result of a stimulus. For example: invertebrate animals move away from a region where it is dry, in case they dry out as well.

In plants the response is slow, by uneven growth (see page 95).

In animals response is rapid, as animals can move part of themselves, or their whole body away from danger.

Choice chamber experiment. Animals such as woodlice are put into the chamber through the centre hole. They move to side II which is moist. This helps them to survive (see page 25).

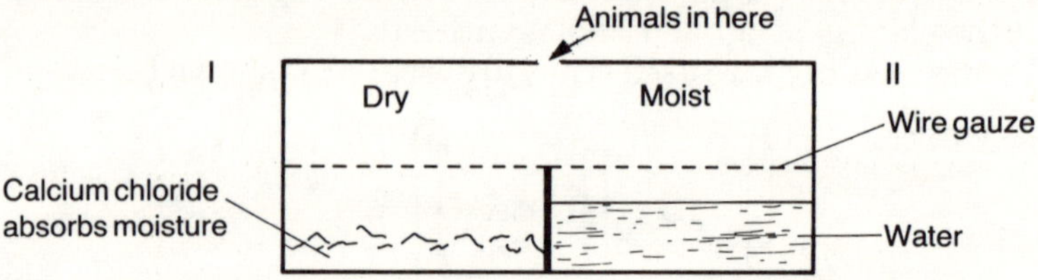

Voluntary and reflex actions

K

VOLUNTARY actions involve thought. The fore-brain or cerebrum is used. Examples are speech and memory.

Impulses from our sense organs (e.g. eye and ear) pass to the brain where they are interpreted and impulses for a response sent out. Voluntary actions depend upon comparing our actions to previous experience. This takes a few moments.

K

REFLEX actions do not involve thought. They happen automatically. The impulses are sent only to the spinal cord or to the lower parts of the brain (cerebellum or medulla).

An example is the iris reflex (where the pupil of the eye is made small in bright light, and wide in the dark, page 99). Another is the withdrawal reflex (see page 97).

Reflex actions allow us to respond quickly to danger and avoid its effects. We don't need to think about our response.

A reflex arc

A REFLEX ARC is the pathway along which nervous impulses pass which are involved in a reflex action. For example, if the hand accidentally touches a hot object, it is immediately withdrawn. This is a reflex action (the withdrawal reflex). The parts of this reflex arc are:

1 Stimulus – this is the heat from the hot object.
2 Receptor – receives the stimulus. The nerve endings in the skin receive the stimulus and nervous impulses are set up in the nerve fibre.
3 Sensory nerve fibre – this takes the impulse from the receptor to the spinal cord.
4 Coordinating nerve fibre – the grey matter of the spinal cord takes the impulse from the sensory nerve and passes on a message to the motor nerve fibre.
5 Motor nerve fibre – takes impulses from the spinal cord to the effector.
6 Effector – responds to the message. In this case the effector is the muscle which contracts.
7 Response – the end result. The hand is moved away.

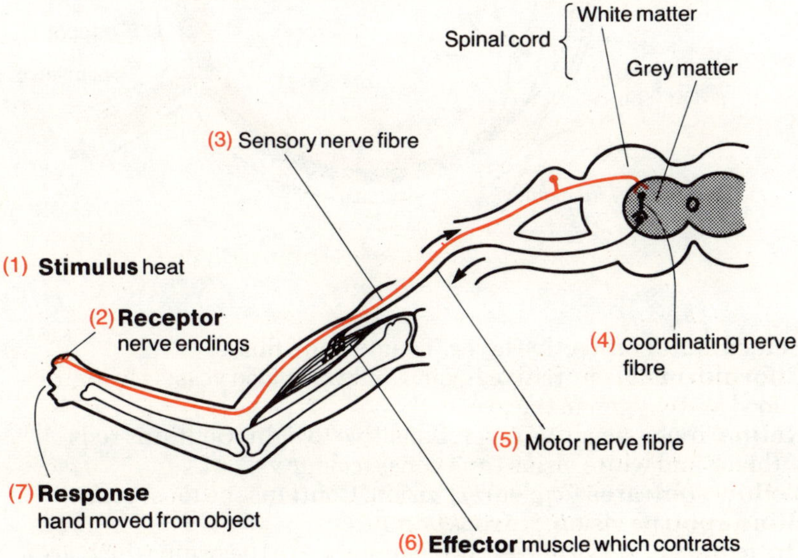

Sense organs

K

SENSE ORGANS are groups of receptor cells sensitive to particular stimuli. The stimuli are various forms of energy which set up nervous impulses which are sent to the brain.

Sense organs	*Sensitive to:*
Eye	light (black and white and colours)
Ear	sound, and important in balance
Nose	chemicals
Tongue	chemicals, touch, temperature, pain
Skin	touch, temperature, pain

The eye

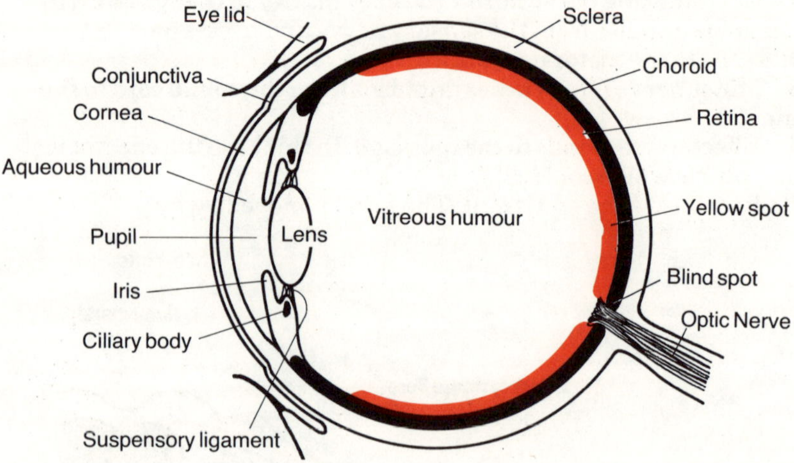

Sclera: outer protective layer. Tough white fibres.
Choroid: middle nutritive layer. Contains blood vessels bringing food and oxygen to the eye.
Retina: inner sensitive layer. Sensitive to light. Contains rods (black and white vision) and cones (colour vision).
Yellow spot: area for clearest vision. Contains the most cones.
Blind spot: no vision. No rods or cones.
Optic nerve: nerve fibres taking impulses to the brain which are interpreted as sight.

Eyelids: protect the eye. Blinking cleans the eye surface.
Conjunctiva: transparent protective cover.
Cornea: transparent extension of the sclera.
Aqueous and **vitreous humours:** help to keep the eyeball in shape. Help in focusing the light.
Pupil: hole through which light enters the eye. Small in bright light (to restrict light so it doesn't damage the retina). Large in dim light (so that as much light as is available can enter the eye).
Iris: controls the size of the pupil (IRIS or PUPIL REFLEX). This occurs automatically as the light intensity changes.

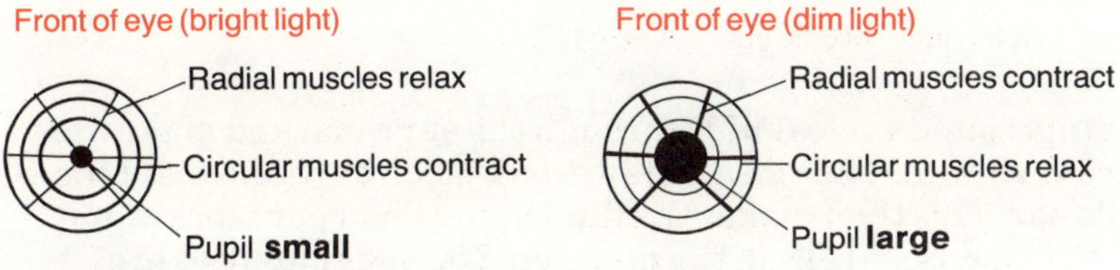

Lens: focuses light to form an image on the retina.
Ciliary muscles and **suspensory ligaments** change the shape of the lens to focus near and distant objects. This change of shape is called ACCOMMODATION.

Near object
e.g. reading a book. Lens becomes fatter.

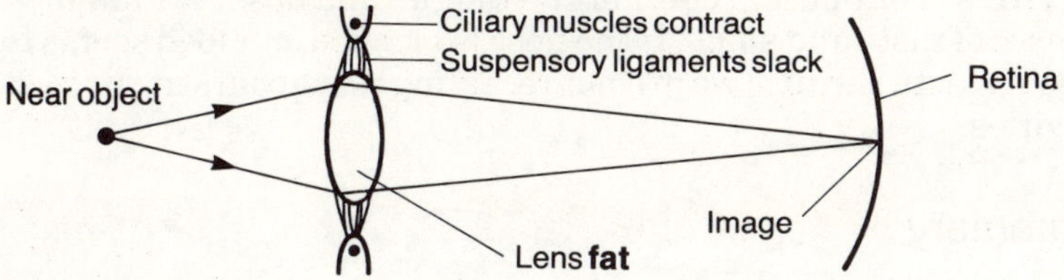

Distant object
e.g. looking outside at a tree. Lens becomes thinner.

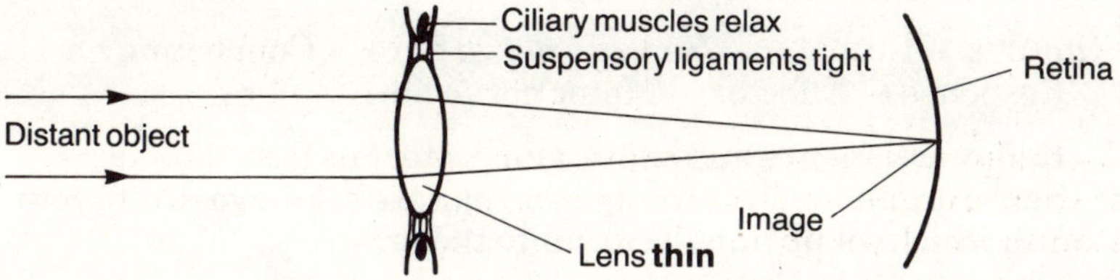

The ear

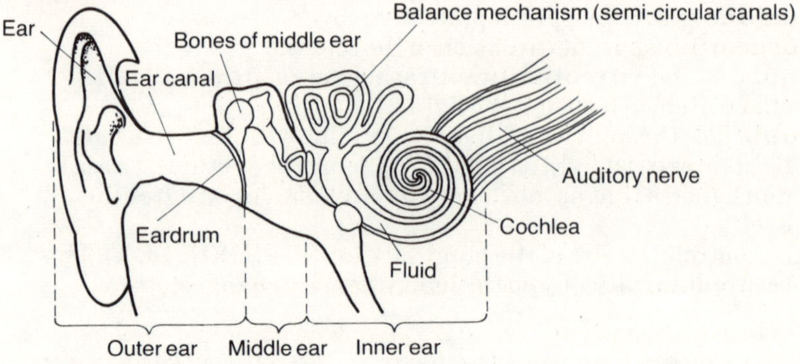

Hearing: sound waves enter through the ear canal and make the eardrum vibrate. The vibrations are passed on by the bones of the middle ear. This then makes the fluid in the inner ear vibrate and this affects sensory cells in the **cochlea**. Nervous impulses are set up which correspond to the original sounds. The impulses are sent along the auditory nerve to the brain where they are interpreted, and we hear.

Taste and smell

Taste buds on the tongue are sensitive to chemicals in our food. Vapours from our food are also sensed by cells in the **nasal cavities** of our nose. Together the tongue and nose give us our senses of taste and smell. If the nose is blocked, our food seems to lose its taste because we are not receiving its vapours in our nasal cavities.

Summary

1 All organisms respond to stimuli.
2 Phototropism is a plant's bending response to unilateral light.
3 In mammals the response pathway is:

Stimulus → Receptor → Sensory nerve fibre → Connecting nerve
Response ← Effector ← Motor nerve fibre ← fibre

4 Reflex actions are automatic and protect us from danger.
5 Sense organs such as the eye and ear are sensitive to different stimuli which set up impulses sent to the brain.

19 Skeletons and movement

Aims of the chapter

After studying this chapter you should be able to:

1 List the functions of plant and animal supporting tissues.
2 Distinguish between different supporting tissues.
3 Describe the elbow joint and movement of the arm.
4 List the importance of other joints and muscles.

Supporting tissues in a flowering plant

Supporting tissues provide **shape** and **form** to the body.

1 The **root** has a central core of xylem which is made up of long vessels which are thickened with lignin. The core helps the root to push its way down into the soil (see also page 77).

2 The **stem** has a ring of vascular bundles each of which has an area of xylem on its inside. The ring arrangement gives strength to the stem and allows it to bend with the wind without breaking (see also page 77).

In young stems the cells of the cortex are swollen with water (turgid) and this helps to keep the stem erect. If water loss exceeds water intake then these cells lose water, become flaccid and the stem wilts.

In older plants the xylem increases in size to make the main part of the wood of the trunk.

3 In **leaves** the leaf stalk, midrib and veins contain xylem which hold up and support the other tissues.

The mammalian skeleton

Functions:

1 **Support:** framework which gives shape and form to the body and to which body organs are attached.

2 **Protection:** bone is hard. Skull protects the brain. Rib cage protects the lungs and heart. Backbone (vertebrae) protects the spinal cord.

3 **Blood formation:** red bone marrow inside certain bones produces red blood cells and phagocytes.

4 **Locomotion:** bones are jointed and provide attachment for muscles which contract to cause movement.

The structure of a synovial joint

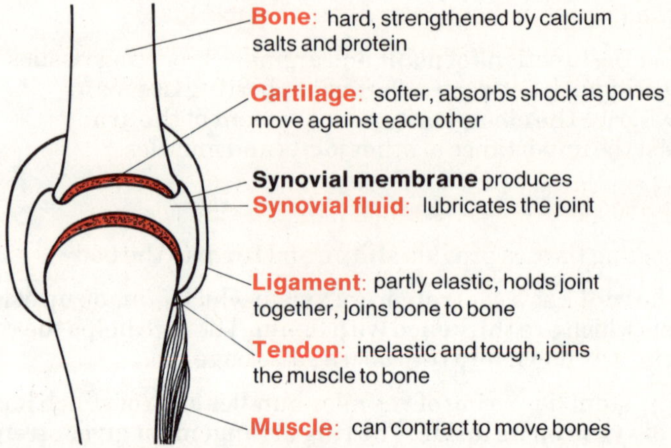

The elbow joint and movement of the arm

K

ANTAGONISTIC MUSCLES are muscles which are arranged in pairs and which have opposite effects when they contract. Muscles cause movement as they contract. As muscles contract they pull closer together the bones to which they are attached.

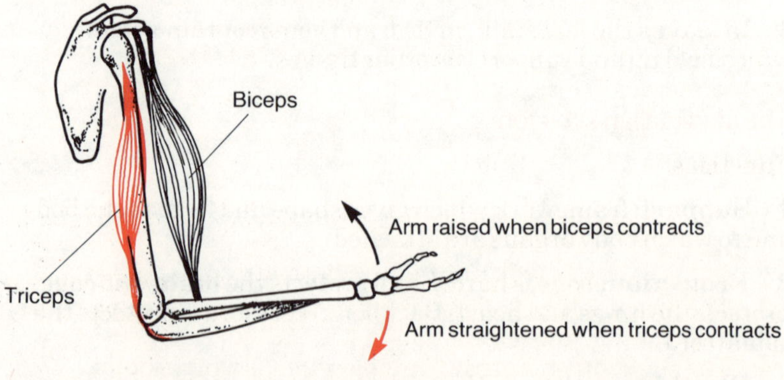

When one muscle of an antagonistic pair contracts, the other one relaxes. So when the arm is raised as the biceps contracts, the triceps relaxes. When the arm is lowered as the triceps contracts, the biceps relaxes.

Kinds of joints

Hinge joints: movement in one direction, up and down. For example, elbow and knee joints.
Ball and socket joints: movement in two directions. Bones can swing around. For example, shoulder and hip joints.

Muscles and movement

Skeletal muscles: muscles attached to the skeleton. A muscle is attached by tendons to bones. When the muscle contracts it pulls the two bones closer to each other. At the same time the other muscle in the antagonistic pair relaxes. Locomotion (movement of the whole body) is bought about by the coordinated contraction and relaxation of muscles. The muscles are effectors which respond to nervous impulses.

Other muscles

1 *Circles* of muscle are found at the exit of the stomach, the anus and the bladder. These keep back the contents of these organs until the ring of muscles is relaxed.
2 *Peristalsis* in the gut is under the control of antagonistic muscles. There are circular and longitudinal muscles in the gut wall and these contract and relax alternately to push the food along the gut.
3 The *iris* of the eye has opposing (antagonistic) circular and radial muscles (see page 99). The pupil is made smaller when the circular muscles contract (and radial muscles relax). The pupil is widened when the radial muscles contract (and the circular muscles relax).

Summary

1 Plants and animals need supporting tissues to provide shape and form.
2 The main supporting tissues are xylem, bone and cartilage.
3 The mammalian skeleton is important for support, protection, blood formation and locomotion.
4 The skeleton is jointed. The elbow joint is a hinge joint.
5 Muscles occur in antagonistic pairs which contract to bring about movement. The other muscle in a pair relaxes.
6 The arm is raised (biceps contracts, triceps relaxes), and lowered (triceps contracts, biceps relaxes).

20 Hormones and coordination

Aims of the chapter

After studying this chapter you should be able to:

1 Give examples of the action of natural and artificial plant and animal hormones.

K 2 Define homeostasis and give examples.

3 Describe temperature regulation and functions of the skin.

Plant hormones

K HORMONES are chemical substances produced in one part of an organism which have an effect on another part.

Natural plant hormones: Auxin (see page 95) is a growth hormone produced in the tip of the shoot. It causes extra growth of the side of the shoot farthest from the light, so causing the shoot to grow towards the light.

Flowering hormone is also made in the shoot tip. It controls the time when buds and flowers will grow.

Artificial plant hormones

1 Rooting hormone powders. Cuttings are dipped in the powder which makes roots grow more easily. (This can be shown by doing a controlled experiment.)

2 Weedkillers may contain growth hormones which make the weeds grow very large and then die.

3 Fruit production can be encouraged by spraying with hormones. Fruit ripening can also be controlled by hormones.

Nervous and hormonal control

The nervous system and hormones both control the actions of the body, but there are several differences:

Nervous system	*Hormonal system*
1 Nervous impulses (electrical)	1 Hormones (chemicals)
2 Rapid transfer	2 Slower transfer
3 Along nerves	3 In the blood
4 Direct to one place e.g. a muscle	4 All over the body, but with certain effects
5 Short-term effect, e.g. response to danger	5 Long-term effect, e.g. increasing growth

The hormonal system

Hormones are produced in special glands, the ENDOCRINE glands, which pass the hormones directly into the blood. The hormones are then taken around the body.

The **pancreas** which lies below the stomach produces INSULIN (see page 106). (*Note.* It also produces pancreatic juice which goes by a duct to the duodenum; this is not a hormone.)

The **adrenal** glands which lie on top of the kidneys produce ADRENALINE. This prepares the body for 'fight or flight'. In response to a dangerous situation adrenaline is produced which causes:

1 An increase in the breathing and heart rates (to increase oxygen supply).

2 More glucose to be released into the blood.

3 Blood vessels to the muscles to be widened to bring them more oxygen and food for greater activity.

The reproductive organs produce SEX HORMONES. OESTROGEN is produced by the **ovary** and PROGESTERONE is produced by the ovary, and the placenta during pregnancy. These hormones cause the development of female secondary sexual characteristics (such as breasts and egg production), and the changes in the menstrual cycle (see page 115).

TESTOSTERONE is produced by the **testis** and it causes the development of male secondary sexual characteristics (such as muscular development and sperm production).

Artificial animal hormones

Contraceptive pills contain oestrogen and progesterone which are taken to *stop* eggs from being released from the ovary. The woman still has monthly bleeding.

Fertility drugs are used to *increase* egg production in infertile women. They are also used in preparation for removing eggs from the ovary prior to fertilization outside the body (IVF = **in vitro f**ertilization). This is done for producing 'test tube' babies. The fertilized eggs are returned to the woman's uterus for them to grow.

At the menopause (when periods stop) a doctor may give oestrogen and progesterone to replace the woman's hormones.

Homeostasis

K

HOMEOSTASIS is the keeping of constant conditions in the body. Body cells can only function properly when, for example, the temperature, glucose and water in the blood are kept within narrow limits.

Changes in the amount of heat or important substances in the blood set off activities which return them to the normal amounts. Some examples are:

1 The control of food concentration by the liver.

2 The lowering of glucose level by insulin.

3 Water balance (osmo-regulation) by the kidney.

1 **The liver.** Food is brought to the liver in the hepatic portal vein from the intestine. Excess glucose is changed to glycogen, and excess amino acids are broken down to give urea and a carbohydrate used for energy. Excess fats are sent for storage under the skin. So the food concentration of the blood is lowered.

Between meals when too little food may be in the blood, the liver changes glycogen to glucose. Fat stores may also be sent to the liver to be made into an energy source. So the food concentration of the blood is raised.

K

2 **Insulin** lowers the amount of glucose in the blood, by changing it to glycogen. People who do not produce enough insulin suffer from **sugar diabetes**. The blood contains too much glucose and some is excreted in the urine.

3 **Osmo-regulation.** Water is lost from the lungs (as water vapour) and from the skin (by the evaporation of sweat) and balanced by the loss from the kidneys (in the urine).

On a hot day

Lungs – lose more water (as warm air can hold more vapour).
Skin – loses more water (as sweating cools the body).
Urine – less in amount, and more concentrated (darker), to conserve water in the body.

On a cold day

Lungs – lose less water (as cold air holds less water vapour).
Skin – loses less water (as sweating reduced to keep warm).
Urine – more in amount, and less concentrated (lighter), as extra water has to be got rid of.

Temperature regulation and the skin

V.S. of human skin

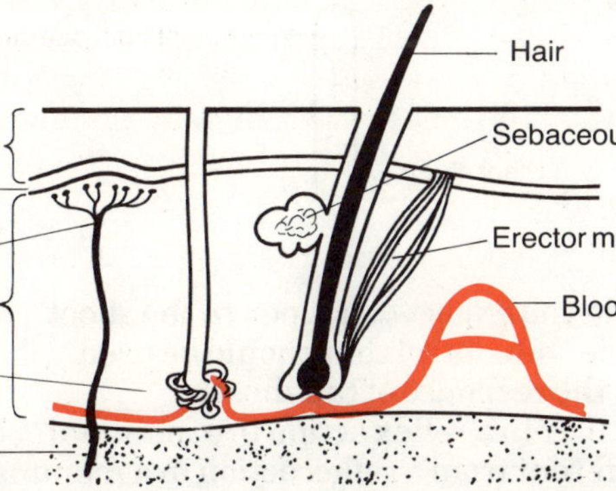

K

Cooling the body

1 Blood capillaries in the skin widen (VASODILATION) to bring more blood to the skin, and heat is lost by radiation.

2 More sweat is produced which takes latent heat from the body as it evaporates, and so cools it.

3 Hairs lie flat (as erector muscles relax). Less air near the body, so evaporation occurs rapidly.

Cooled blood circulates and cools the body.

K

Warming the body

1 Blood capillaries in the skin narrow (VASOCONSTRICTION) so less blood is brought to the skin, so less heat is lost.

2 Sweating decreases, so less heat is lost.

3 Hairs stand up ('goose pimples') as erector muscles contract. Layer of moist air trapped which reduces sweating.

Warmer blood circulates and warms the body.

Summary

1 Hormones are chemicals which have an affect on the body.

2 Important plant hormones are auxin (growth hormone) and rooting hormone.

3 Adrenaline is the 'fight or flight' hormone.

4 Insulin lowers the glucose level of the blood.

5 The kidneys balance the water loss in the body.

6 The skin regulates the temperature of the body.

Questions on Chapters 18–20

70 The diagram below shows an experiment to find the effect on a plant of light from one direction.

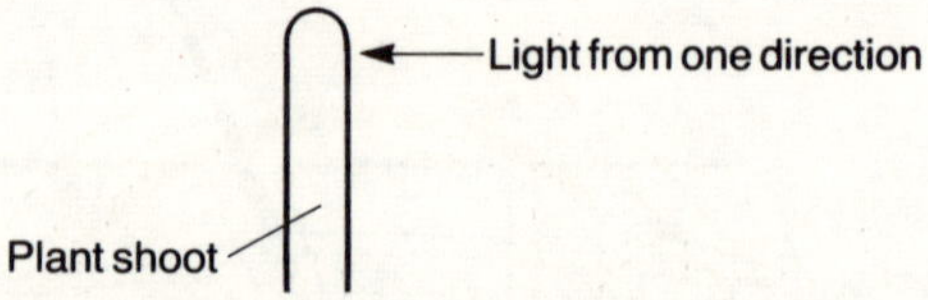

(a) What do you expect to happen to the shoot?
(b) Describe the control that should be used.
(c) How is the response of the shoot:
  (i) similar to a reflex action in a mammal? (Two ways.)
  (ii) different from a reflex action in a mammal? (Two ways.)

71 The diagram below shows the parts of a spinal reflex arc.

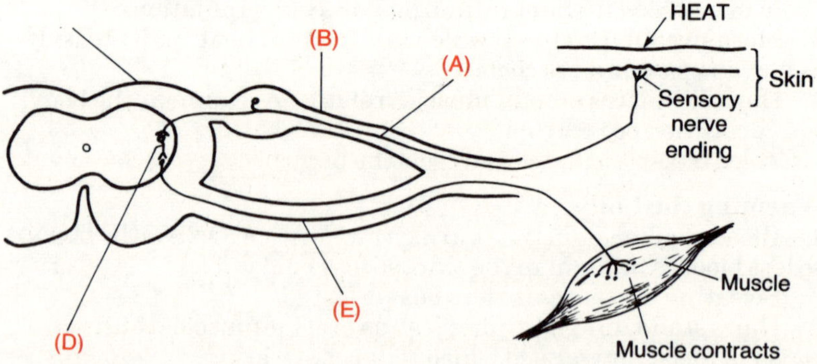

(a) Correctly match an item on the left with items I, II, III and IV on the right.

(i) Heat	I Receptor
(ii) Sensory nerve endings	II Effector
(iii) The muscle	III Response
(iv) The muscle contraction	IV Stimulus

(b) Name the structures (A)–(E)
(c) Of what importance are spinal reflex actions to the body?
(d) Give TWO ways in which a spinal reflex action differs from a voluntary action.

72 The diagram below shows a section through the human eye.

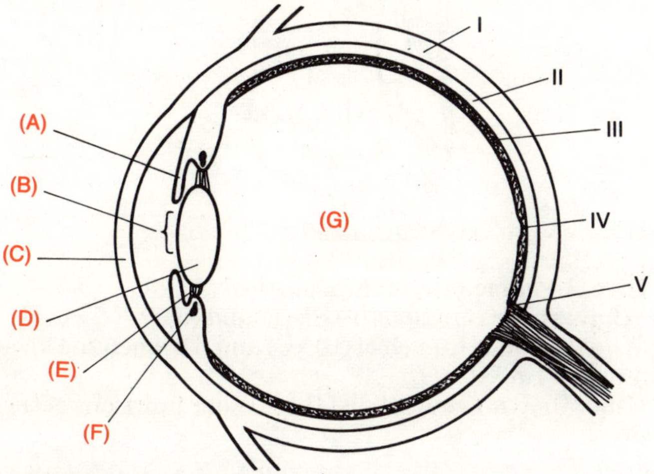

(a) Label the structures (A)–(G)
(b) (i) What change occurs in (B) when in dim light?
(ii) How is this brought about?
(c) (i) How do (E) and (F) make (D) thicker (fatter)?
(ii) Under what conditions would this be necessary?
(d) (i) Give the names of the structures I–V and write one sentence about each one to describe its function.
(ii) The image formed on the retina is inverted ('upside down'). How is it that we see it the right way up?

Questions **73–76** have five possible answers: (A), (B), (C), (D) or (E). For each question choose the best response. A response may be used once, more than once or not at all.

(A) Adrenaline (B) Insulin (C) Testosterone
(D) Pancreatic juice (E) Auxin

Which of the above:
73 Is produced in the pancreas and lowers the amount of glucose in the blood? ____
74 Controls the development of male secondary sexual characteristics? ____
75 Is produced by plants? ____
76 Causes increase in heart rate under stress? ____

77 The diagram below shows a section through the elbow joint, together with the associated structures.

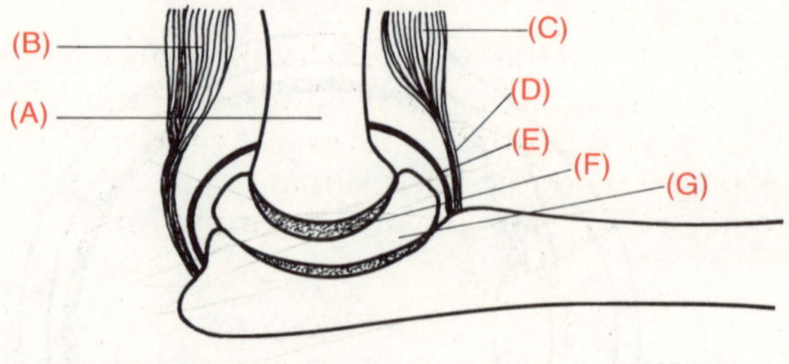

(a) Give the names of the parts labelled (A)–(G).
(b) What are the functions of (E), (F) and (G)?
(c) What happens to each of (B), (C) and (D) when the lower part of the arm is raised?
(d) Other than movement, list three other functions of the skeleton.

78 The diagram below shows a section of the human skin.

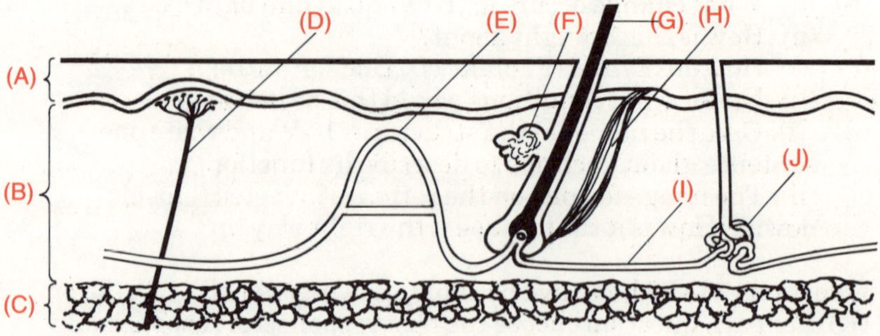

(a) Give the names of the parts labelled (A)–(J).
(b) On a HOT day what happens to each of the structures (E), (H), (I) and (J), and how does each one help to cool the body?
(c) Which of the structures are responsible for:
  (i) sensing changes in the environment? ____
  (ii) providing an insulation against the cold? ____
  (iii) 'goose pimples'? ____.
  (iv) excretion of wastes? ____

Now turn to page 150 for answers 70–78.

21 Growth and reproduction

Aims of the chapter

After studying this chapter you should be able to:

1 Describe growth and the stages in the life cycle.
2 Describe asexual reproduction in flowering plants.
3 Distinguish between asexual and sexual reproduction.
4 Describe sexual reproduction in humans.
5 Describe sexual reproduction in flowering plants.

Growth and life cycles

GROWTH is the irreversible increase in size with the making of new protoplasm. It usually includes the making of new cells by cell division. It can be measured by:

1 Increase in mass, e.g. by measuring mass of guinea pigs.
2 Increase in dry mass (organisms which have been dried by driving off all the water), e.g. by measuring seedlings.
3 Increase in length, e.g. of human height, length of an insect, length of root or stem.

DEVELOPMENT is the increase in complexity of an organism; how it changes throughout its life. The cells that are formed take on different functions. **Maturity** is the stage of development when the organism can reproduce.

REPRODUCTION is the making of new organisms similar to the parent or parents. It is part of the whole life cycle which ensures the continuity of life.

LIFE CYCLE: the time from the formation of an organism (plant or animal) until the organism is mature and can make more organisms. A life cycle with sexual reproduction is shown below.

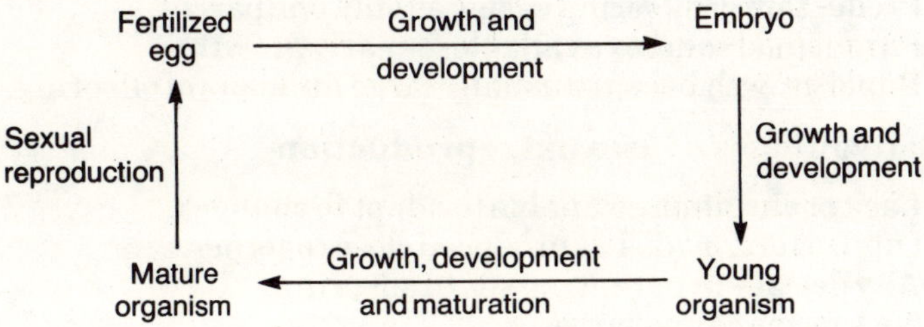

Asexual reproduction

K

ASEXUAL REPRODUCTION produces offspring with identical genes to the parent plant. Only one parent. No gametes.

***Amoeba* and bacteria:** unicellular organisms grow and split into two indentical halves **(binary fission)**.

Fungi: reproduce asexually by forming **spores** in spore cases. Spores are shot out and form new threads (hyphae).

Flowering plants. Part of the plant can grow into whole new plants identical with the parent. For example:

1 *Stem cuttings*. A piece of stem, e.g. of Busy Lizzie, is cut from the plant and most of the leaves removed (to reduce water loss). This is pushed into the soil and it forms roots. It can make a whole new plant. Root production can be improved by first dipping it in rooting hormone.
2 *Leaf cuttings,* e.g. African violet and Begonia. Leaf pieces kept moist will form roots and new plants.
3 *Runners,* e.g. Strawberry. A bud grows out to form a stem which grows (runs) along the ground. At certain places it forms roots in the soil and new separate plants can grow.
4 *Potato*. Each of the 'eyes' on a potato can sprout and if put into the soil can grow to make a new plant. Each plant makes stems into the soil which swell to make more potatoes.
5 *Tissue culture*. Single cells, e.g. of orchid, are grown in food material to form whole new plants like the parent.
6 *Bulbs,* e.g. onion and daffodil, have food stores in swollen leaves. This food allows young buds to grow out to make new plants. After flowering, the plants make more new bulbs.

Advantages of asexual reproduction

1 Offspring identical with parent so new varieties can be kept.
2 Production of offspring easier as only one parent.
3 Parent food sources available for early growth.
4 Rapid growth because usually large numbers of offspring.

Disadvantages of asexual reproduction

1 Lack of variation so unable to adapt to changes.
2 Often overcrowded as produced close together.
3 Any disease in parent passes to offspring.
4 New forms cannot arise.

Sexual reproduction

SEXUAL REPRODUCTION involves the production of special cells, the **gametes**.

Male gametes	**Female gametes**
Pollen (flowering plant) Sperm (human)	Egg in ovule (flowering plant) Egg in ovary (human)
Smaller in size	Larger in size
Dispersed or can swim	Not dispersed nor able to swim
Needs food supply	Contains some food store
Large numbers	Small numbers

Gametes are formed by reduction division (or MEIOSIS) which halves the number of chromosomes. In the process of meiosis different chromosomes go into different gametes so that there is a large variety of gametes.

Gametes fuse in pairs (FERTILIZATION) to make the fertilized egg (zygote) which develops into the **embryo**. In a flowering plant this occurs in the seed and the embryo germinates to make the young plant (the seedling). In humans the embryo develops in the uterus, and is later born.

Advantages of sexual reproduction

1. Variety of organisms allows spread into new areas.
2. Greater chance of survival in changing conditions.
3. Better forms (for crops and breeding) can arise.

Disadvantages of sexual reproduction

1. Need two parents and process is more complicated.
2. New variety cannot be kept pure for breeding.
3. Usually takes longer for young to develop.

Asexual reproduction	*Sexual reproduction*
1 One parent	1 Two parents
2 Mitosis forms offspring which then grow	2 Meiosis forms gametes which fuse in pairs and then grow
3 Identical genes with parent	3 Different mix of genes
4 Allows rapid growth in good conditions	4 Variety allows survival in new or poor conditions
5 Use it to keep new varieties the same	5 Use it to develop new varieties in the first place

Male reproductive system

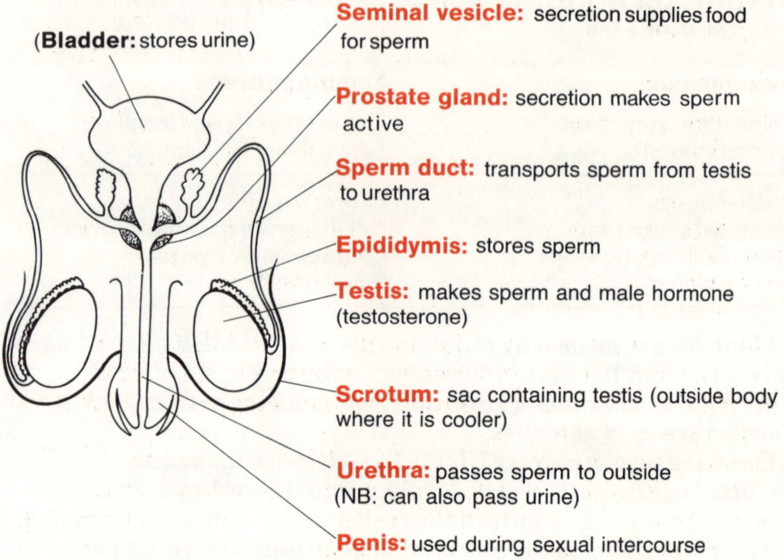

Female reproductive system

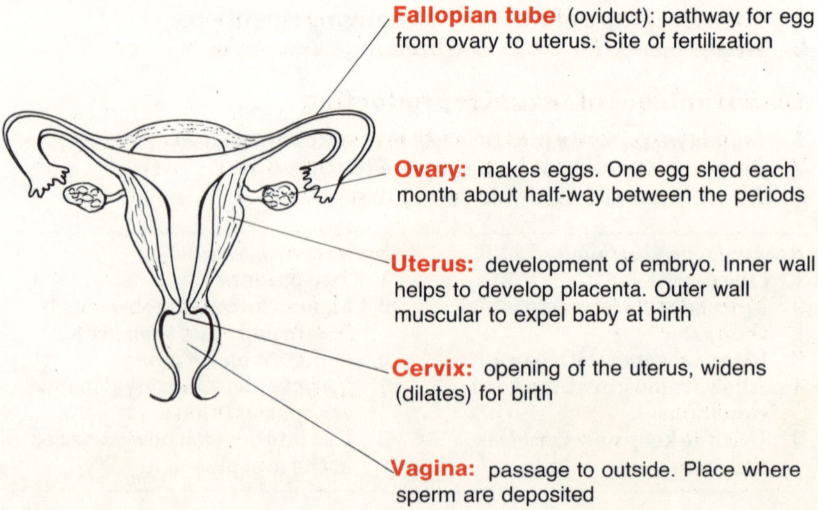

The menstrual cycle

The MENSTRUAL CYCLE is the repeated cycle of events which occur each month from **puberty** (when egg production begins) to **menopause** (when egg production stops). It is under the control of the female hormones, oestrogen and progesterone.

Days 1–5: bleeding from the uterus (called the 'period').
Days 6–13: uterus wall repaired, egg grows in the ovary.
Days 13–15: release of mature egg from ovary **(ovulation)**.
Days 15–25: uterus wall thickened by extra blood.
Days 26–28: uterus wall begins to break down.
Then the cycle is repeated from Day 1.

(NB. The above cycle is a regular 28-day cycle. But many women have irregular cycles and cannot predict the time of the release of the mature egg. If intercourse occurs when there is a mature egg, then the woman may become pregnant. When using the contraceptive pill, no eggs are produced but there is still monthly bleeding.)

Insemination, fertilization and implantation

INSEMINATION is the introduction of sperm into the vagina. This is usually by the natural process of sexual intercourse when the penis is inserted into the vagina. Stimulation causes ejaculation and release of sperm. A doctor can also insert sperm by artificial insemination.

K FERTILIZATION is the fusing of a sperm with an egg. It usually occurs in the Fallopian tubes. After fertilization the zygote divides into a ball of cells which continues its way down the Fallopian tube to the uterus. Doctors can remove eggs from the ovary if a woman has blocked tubes. In this case fertilization can occur outside the body (IVF = In Vitro Fertilization). Babies begun in this way are sometimes called 'test tube' babies, but the ball of cells has to be put into the woman's uterus to continue its development.

K IMPLANTATION is the fixing of the ball of cells into the inner wall of the uterus. The uterus wall and the tissues of the embryo form the placenta, and the woman is pregnant. During pregnancy the inner wall is not shed; the woman does not have any periods.

Development of the foetus

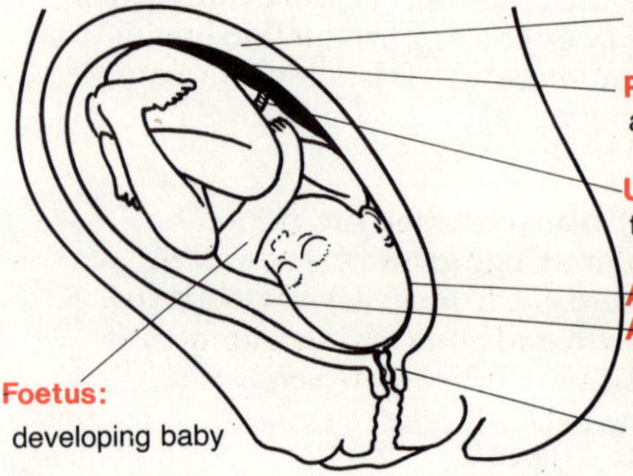

To the foetus in umbilical vein from the placenta:
Oxygen, Food, Vitamins and Minerals, Antibodies
(NB. Danger: viruses, e.g. German measles and AIDS, and drugs can also get to the foetus and harm it.)

From the foetus in umbilical artery to the placenta:
Wastes: Carbon dioxide, Urea

Birth

After about 9 months development **(gestation)** the baby is ready to be born:

1 The muscles of the uterus begin to contract.
2 The cervix at the exit of the uterus becomes wider.
3 The amnion bursts to release the amniotic fluid.
4 The contractions become closer together and more powerful.
5 The contractions push the baby out through the vagina.
6 Then the placenta is expelled as the 'afterbirth'. The umbilical cord attached to the baby is tied in two places and cut in between. The baby has to breathe for itself.

Parental care

In fish, fertilization is **external** and many eggs and sperm are produced. There is rarely any parental care.

In mammals, fertilization is **internal** and fewer eggs and sperm are produced. The young develops in the female. When born it is fed on breast milk and protected as it grows up.

Flower structure and pollination: Wallflower

Functions of the flower parts

Sepals: usually small and green. Protect other parts in the bud. Support the petals.
Petals: large and brightly coloured. Attract insects by their colour and scent.
Nectaries: make nectar which attracts insects.
Stamens: anther produces pollen, filament holds up anther.
Carpel: stigma receives pollen, style directs pollen tube to the ovary, ovary produces ovules (containing female gamete).

L.S. of wallflower

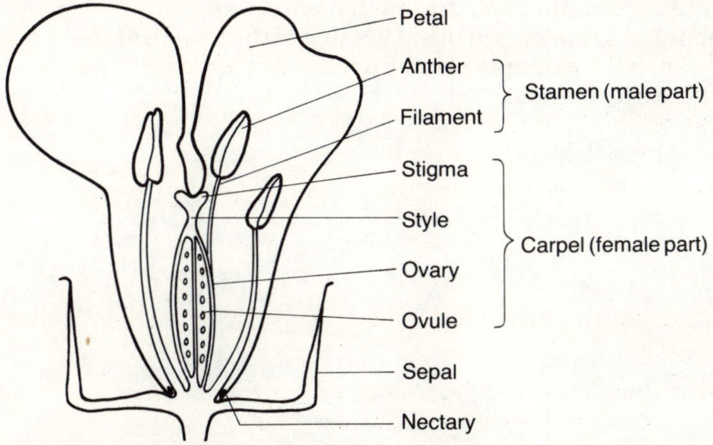

Pollination of wallflower

1 Insects are attracted to the flowers by their red and yellow petals, scent and nectar.
2 The anthers produce pollen which is like yellow dust as the anthers open.
3 The insect gets some of the ripe pollen on its underside.
4 The insect visits another wallflower and pollen is transferred to the sticky stigma.
5 The pollen grains split open and pollen tubes grow down the style towards the ovary.
(NB. If the insect goes to a flower of a different species, the pollen just dies.)

Flower structure and pollination: French bean

(NB. Check with your teacher to see whether wallflower or French bean is required for your examination group.)

The French bean is a variety of *Phaseolus vulgaris*. (Other varieties are string bean, green bean and kidney bean.) It is a leguminous plant which produces pods (the fruits), and its roots have nodules with nitrogen-fixing bacteria.

Sepals: small and green, protect the bud.
Petals: usually coloured in cultivated forms. The large standard petal stands upright. Two wing petals are at the side and two keel petals surround the carpel inside the flower.
Stamens: the filaments of 9 of the stamens are joined together to make a trough. The 10th stamen is free.
Carpel: this is elongated and lies inside the trough of the stamens. It contains about 5 ovules. At the end of the style is the sticky stigma.

L.S. of French bean.

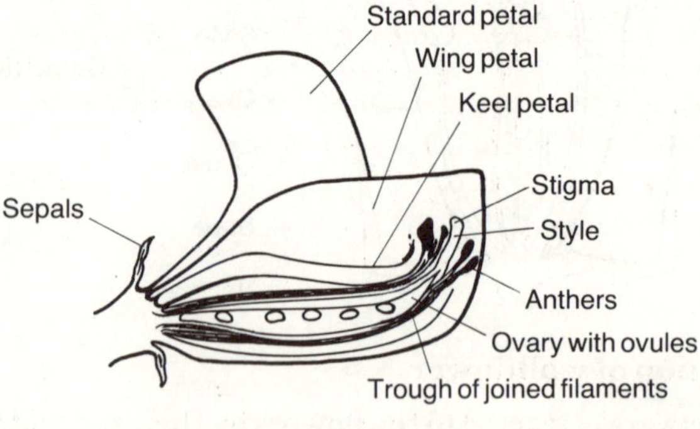

Pollination of the French bean

1 Very small insects are able to crawl around inside the tube made by the keel petals.
2 Pollen is shed from the stamens and this either falls directly onto the stigma (if it is ripe) or onto the insects.
3 If the insects transfer the pollen to the stigma of the same plant this is **self-pollination**. (This is most common.)
4 If the insect leaves the flower and goes to another one and transfers the pollen to that stigma it is **cross-pollination**.

Insect and wind pollination

Insect pollination

1 Large coloured, scented petals attract insects

2 Usually produce nectar

3 Stamens inside flower so insect will touch them

4 Pollen sticky or spiny to attach to insects

5 Stigma sticky. Held inside flower where insect will go

Wind pollination

1 Small or absent petals do not get in the way of pollen

2 Do not produce nectar

3 Filaments long, so anthers hang outside flower

4 Dry, smooth pollen. Light to blow easily in the wind

5 Feathery, sticky stigma hangs outside flower to catch pollen

Fertilization

FERTILIZATION is the fusing of the male nucleus (from the pollen) with the female nucleus (in the ovule).

Self-fertilization follows self-pollination when pollen is transferred from stamen to stigma of the same flower.

Cross-fertilization follows cross-pollination when pollen is transferred from the stamen to the stigma of another flower of the same species. This gives more variation.

The pollen grain splits open on the stigma and a pollen tube grows down the style to the ovary and into the ovule.

The fertilized **ovule** becomes the **seed**.

The **ovary** containing fertilized ovules becomes the **fruit**.

Seed structure and germination, e.g. French bean

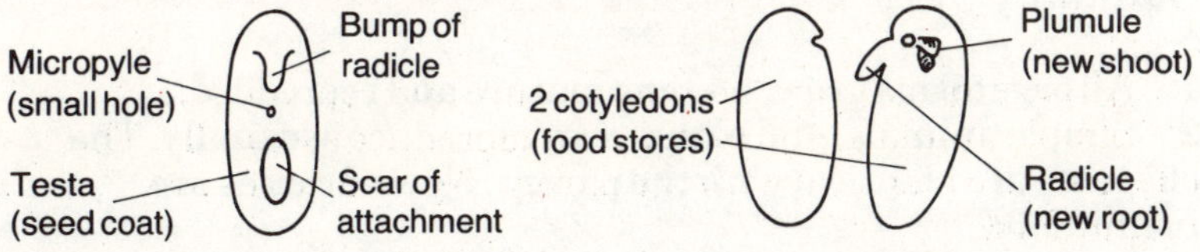

The seed takes up water and swells. The testa splits and the radicle emerges. It grows down into the soil. The part just below the cotyledons grows and takes the cotyledons and the plumule up through the soil. The cotyledons supply food and also turn green to carry out photosynthesis. The plumule then grows out and forms the first true leaves. The cotyledons shrivel and drop off. The seedling continues to grow.

Conditions necessary for germination

Seeds are set up in experiments lacking just one factor. They are compared to a CONTROL with all the factors.

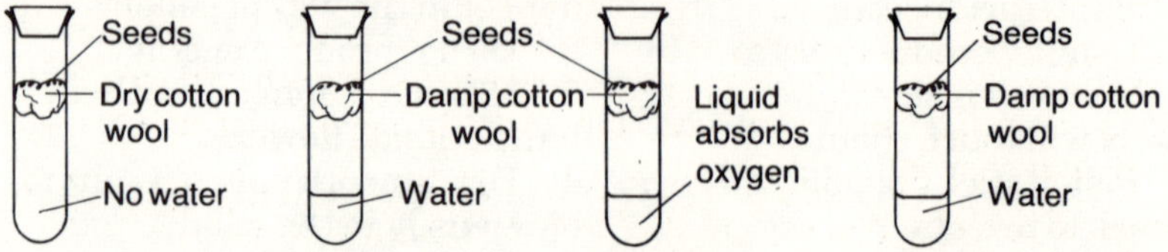

In room	In fridge	In room	In room
No water	No warmth	No oxygen	**Control**
No germination	No germination	No germination	Water, warmth, air
			Germination

Dispersal and colonization

K

DISPERSAL is the scattering of fruits and seeds away from the parent so that they will not be overcrowded. Some will fall onto new undeveloped ground and **colonize** it.

Self-dispersal, e.g. wallflower and wild ancestors of French bean: fruit splits open and seeds are shot out.

Wind dispersal, e.g. dandelion: parachute of fine hairs on the fruit makes it light and easily blown by the wind. In other plants the seed coat (testa) or fruit wall (pericarp) are extended to form wings.

Animal dispersal, e.g. burdock: small hooks on the fruit wall attach to hair on animals and are dispersed.

Summary

1 All organisms grow, become mature and reproduce.
2 Simple animals, and plants, can reproduce asexually. The offspring are identical with the parent. New varieties are maintained.
3 All organisms reproduce sexually. Gametes fuse to give offspring different to the parents. New varieties can arise.
4 In humans, fertilization is followed by implantation and the foetus develops in the uterus with the placenta and amnion.
5 Flower structure is modified for insect or wind pollination.
6 Fertilization follows pollination and leads to seed production:
Ovule → seed Ovary → fruit
7 Germination requires water, warmth and oxygen.

79 (a) List THREE ways in which growth in plants can be measured.
(b) List TWO ways in which growth in plants differs from growth in animals.

80 (a) Divide the following structures into two groups: those concerned with *asexual* reproduction, and the others concerned with *sexual* reproduction.

Stamens, Bulbs, Corms, Ovules, Petals, Runners, Spores, Gametes

(b) Describe ONE advantage to the farmer of asexual reproduction in flowering plants.

81 (a) List THREE differences between insect- and wind-pollinated flowers.
(b) How could a gardener:
(i) Prevent cross-pollination of a flower?
(ii) Prevent self-pollination of a flower?

82 The diagram below shows the male reproductive system and part of the excretory system from the side.

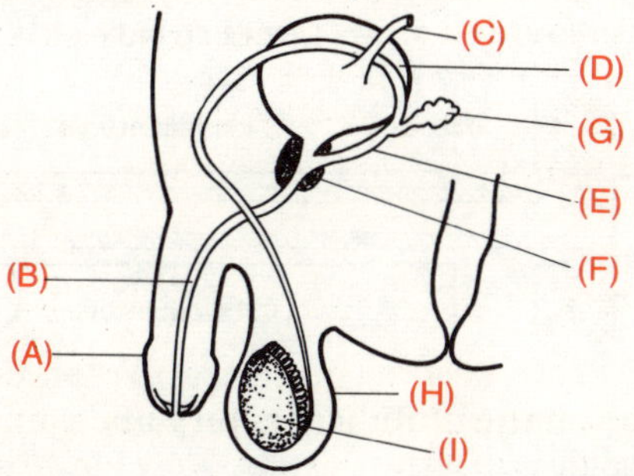

(a) Name each of the structures (A)–(I) and write ONE sentence about each to describe its function.
(b) Which of the structures (A)–(I) are part of the excretory system?
(c) During the operation of male sterilization (vasectomy) part of the tube (E) is cut and removed.
(i) What effect will this have on the transfer of sperm?
(ii) For three to six months after the operation the man might still father a child. How is this possible?

83. The diagram below shows the female reproductive system.

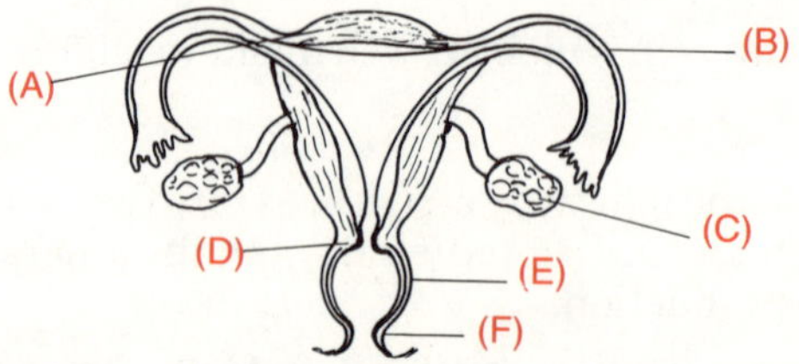

(a) Name each of the structures (A)–(F) and write ONE sentence about each to describe its function.
(b) In the female, the vagina and urethra open separately to the outside. Give TWO reasons why this is important.
(c) What happens to the lining of (A) at the following times?
  (i) During days 1–5 of the menstrual cycle.
  (ii) From days 6–21 of the menstrual cycle.
  (iii) During pregnancy.
(d) List FOUR causes of infertility (of either male or female).

84 The diagram below shows in a simple way how the human foetus is connected to the placenta.

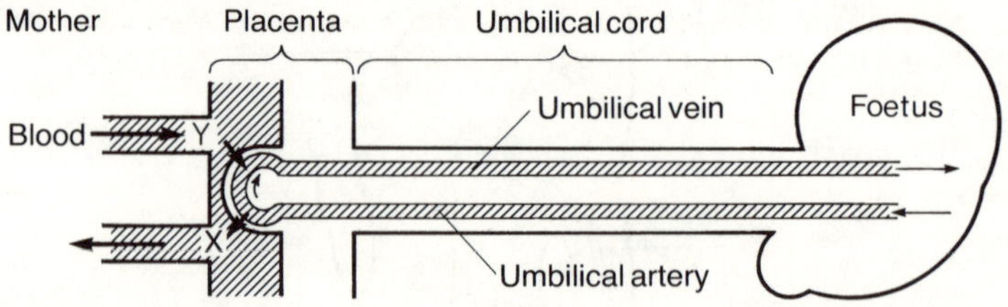

(a) Draw arrows on the umbilical artery and vein to show the direction of blood flow.
(b) (i) Name two substances which pass along pathway **X**.
  (ii) By what process do they pass?
(c) (i) Name TWO useful substances which pass along the pathway marked **Y**.
  (ii) Name TWO harmful substances which might pass along the pathway marked **Y**.
(d) After the baby has been born what happens to:
  (i) The umbilical cord?
  (ii) The placenta?

85 The diagram below shows the parts of an opened bean pod.

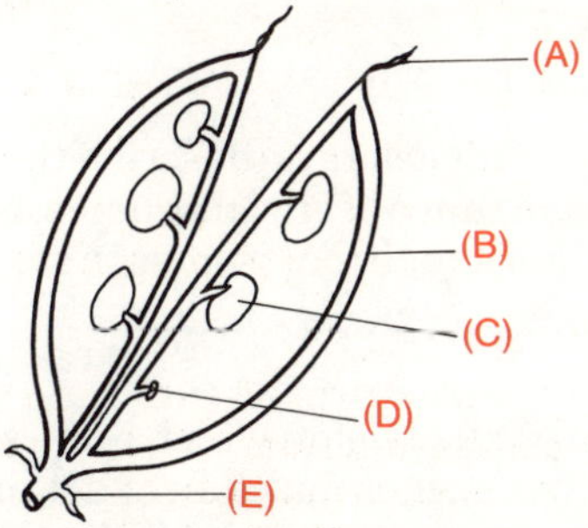

(a) What are the following parts derived (produced) from?
(A) ________________
(B) ________________
(C) ________________
(E) ________________
(b) Suggest a reason for the difference between (C) and (D).
(c) How will the seeds be dispersed from the pod?

86 The diagram below shows the control set-up used to find the conditions necessary for germination.

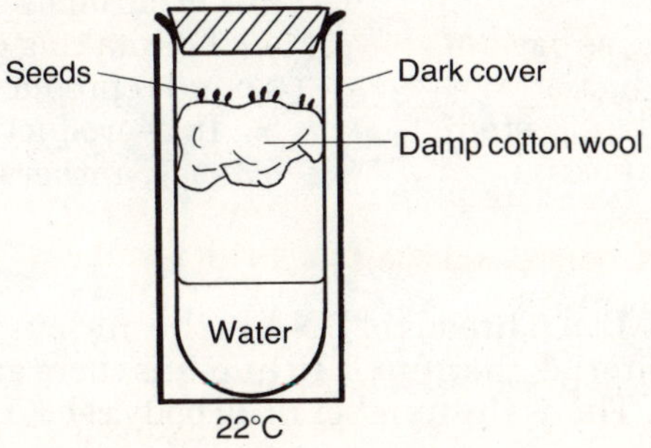

(a) List the FOUR conditions that have been supplied to these seeds.
(b) Why are 8 small seeds used instead of just one large seed?

Now turn to page 152 to check your answers 79–86.

22 Chromosomes and how cells divide

Aims of the chapter

After studying this chapter you should be able to:

1 Define mitosis and meiosis and describe their importance.
2 Describe the importance of chromosomes, genes and alleles.
3 Describe the role of sex chromosomes in sex determination.

Mitosis and meiosis

MITOSIS is ordinary cell division which occurs during growth and repair. The chromosomes make an exact copy of themselves and are divided between two new cells. The new cells have the same number of chromosomes as the parent cell.

MEIOSIS is reduction division which occurs in the making of gametes. The chromosomes make an exact copy of themselves, but the parent cell divides into four new cells so each one has half the original number of chromosomes.
(NB. Fertilization restores the full number of chromosomes.)

Mitosis	*Meiosis*
1 Ordinary cell division	**1** Reduction division
2 Same number chromosomes	**2** Half number chromosomes
3 New cells identical with parent cell	**3** New cells have different mix of chromosomes
4 For growth, as basis of asexual reproduction	**4** For making gametes in sexual reproduction
5 In body cells, e.g. stem and root tips (apices)	**5** In reproductive organs: ovaries, anthers, testes

Chromosomes, genes and alleles

CHROMOSOMES. Threads of DNA in the nucleus which carry the genetic material, the **genes**. In humans there are 23 pairs of chromosomes. This is the number in all body cells. There is a characteristic number of chromosomes for each species.

GENES. The parts of the chromosome which determine inherited characteristics, e.g. the gene for height in sweet peas. The **alleles** are the alternative forms of the gene for a certain characteristic, e.g. for tall and short in sweet peas. Only one allele of each pair is in the gamete.

Sex chromosomes and sex determination

In humans there are 23 pairs of chromosomes. 22 pairs are the ordinary chromosomes and one pair are the sex chromosomes. SEX CHROMOSOMES determine the sex of the person. If the chromosomes are alike (XX) the person is female, if they are different (XY) the person is male.

Father	**Mother**
22 pairs + XY	22 pairs + XX
(Meiosis makes sperm of two kinds)	(Meiosis makes eggs all alike)

Either sperm can fertilize either egg giving these results

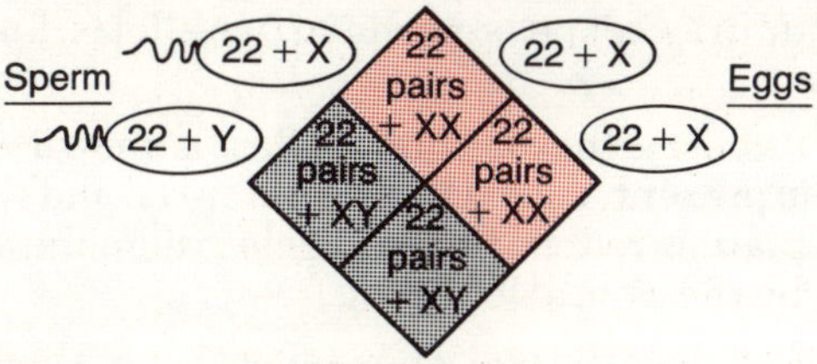

Half offspring (22 pairs + XY) are male and half (22 pairs + XX) are female.

Twins

Identical twins are formed if the fertilized egg divides into two or more parts and each part grows into a baby. Identical twins have the same genes and are either all male or all female.

Non-identical twins are formed if two or more eggs shed from the ovaries are fertilized. They can be either both male or both female, or one male and one female. They do not have identical genes.

Summary

1 Mitosis (ordinary cell division) maintains the same chromosome number as body cells divide.
2 Meiosis (reduction division) halves the chromosome number in gametes prior to fertilization.
3 Chromosomes carry genes which determine characteristics.
4 Female sex chromosomes are XX, male ones are XY.

23 How characteristics are passed on

Aims of the chapter

After studying this chapter you should be able to:

K

1 Define the terms alleles, dominant, recessive, homozygous, heterozygous, phenotype, genotype, complete dominance.
2 Predict the offspring expected from certain parents.
3 Define and give examples of continuous and discontinuous variation.
4 Define and give examples of mutations.

Definitions

ALLELES are the alternative forms of a gene. For example, the gene for colour in sweet peas can have two alleles, for red and for white.

DOMINANT allele is the one whose effect is shown when there is another allele present, e.g. if the alleles for red and white are both present, the plant is red. So the red allele is dominant. It is represented by the capital letter, 'R'.

RECESSIVE allele is the one whose effect is not shown when there is another allele present, e.g. if the alleles for red and white are present, the plant is red. The white allele does not show its effect. It is recessive. It is shown by small 'r'.

HOMOZYGOUS describes an organism which has the same alleles of a particular gene. e.g. homozygous red would be RR, and homozygous white would be rr.

HETEROZYGOUS describes an organism which has different alleles of a particular gene. The heterozygote would be Rr. R (red) is dominant to r (white) so this plant would appear red.

PHENOTYPE is the outward appearance, e.g. red flowered.

GENOTYPE is the internal description of the actual alleles of an organism, e.g. a plant that looks red (phenotype red) could have either the genotype RR or Rr.

COMPLETE dominance is when the homozygote (RR) looks like the heterozygote (Rr). One of the alleles (R in this case) is completely dominant.

INCOMPLETE DOMINANCE is when neither allele is completely dominant. The heterozygote looks different.

Genetic crosses

The crosses you will be given are of three kinds:

Homozygote × Homozygote (RR × RR, rr × rr, RR × rr)
Heterozygote × Heterozygote (Rr × Rr)
Homozygote × Heterozygote (RR × Rr, rr × Rr)

You may be given plants or animals and any kind of characteristic, but the pattern will be like one of the following.

R (red allele) is dominant to r (white allele).

Homozygote × Homozygote

RR × RR
(pure breeding)

Parents RR × RR

Gametes

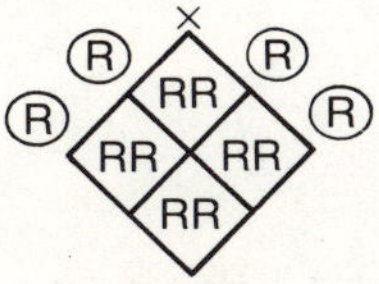

(NB: only one allele of a pair goes into a gamete)

Offspring all RR (red)

rr × rr
(pure breeding)

Parents rr × rr

Gametes

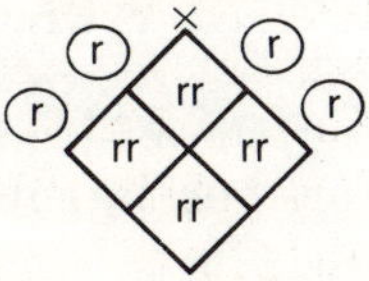

Offsring all rr (white)

RR × rr
(offspring as dominant parent)

Parents RR × rr

Gametes

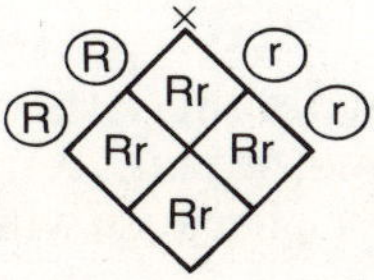

Offspring all Rr (red)

Heterozygote × Heterozygote

Rr × Rr
(gives 3 : 1 dominant : recessive)

Parents Rr × Rr

Gametes

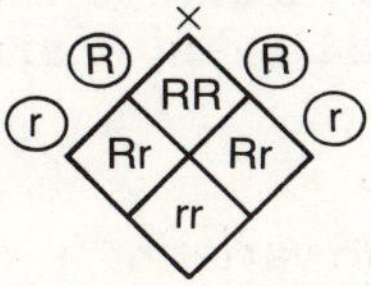

Genotypes RR Rr Rr : rr

Phenotypes 3 red : 1 white

1 of the red plants (RR) will be pure breeding (only gives red).
2 of the red plants (Rr) when crossed with each other, or self-pollinated, will give the same 3 red : 1 white.

Homozygote × Heterozygote

RR × Rr
(half red pure, half red impure)

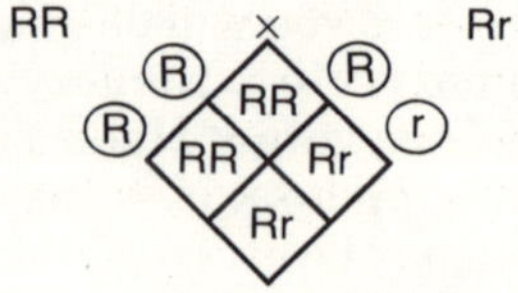

Phenotypes All red
Genotypes RR RR : Rr Rr
½ pure breeding : ½ not pure breeding

rr × Rr

Genotypes Rr Rr : rr rr
Phenotypes ½ red : ½ white

You may be given the offspring (also called the F_1 generation) and asked to work out the genotypes of the parents:

1 Plant A × Plant B→ All red
(red) (white) (F_1 generation)

Red allele is dominant (R) because offspring (Rr) are red. All offspring are red, so plant A must be pure red (RR). Plant B is homozygous (rr), because white is recessive.

2 Plant C × Plant D→ Half red : Half white
(red) (white) (F_1 generation)

Plant C must be heterozygous (Rr) because there are some white offspring. If self-pollinated it would not breed true. D is (rr). Plants A, B and D are all homozygous and would breed true.

Variation

CONTINUOUS variation is when there is a gradual change in a characteristic from one extreme to another, e.g. height or mass in a population. It is controlled by many genes.

DISCONTINUOUS variation is when there are distinct groups with no 'in-betweens', e.g. able or not able to roll the tongue. It is controlled by one gene.

Mutations

MUTATIONS are changes in a gene or chromosome. They can be caused by X rays, radiation, chemicals etc.

An example of a gene mutation is haemophilia.

An example of a chromosome mutation is **Down's syndrome** where the child has an extra chromosome which upsets normal development. Down's syndrome children are more common with older mothers where cell division to produce eggs has not occurred properly, because of ageing.

Genetic engineering

GENETIC ENGINEERING is the changing of genes/alleles. The allele, e.g. for producing insulin, can be removed from the human chromosome and inserted into the circular chromosome of a bacterium. The bacterium will then make insulin. Single genes or alleles have to be transferred. Bacteria are useful because of their simple structure and rapid reproduction.

There are now experiments to transfer alleles for resistance to some diseases into plants. It may also be possible to transfer normal genes to replace damaged ones in humans.

Summary

1 Alleles are different forms of the same gene. Only one allele of a pair goes into a gamete.
2 Dominant alleles show their effect in the heterozygote.
3 Recessive alleles don't show their effect in the heterozygote.
4 For a recessive allele to show it must be homozygous.
5 Homozygotes (RR or rr) breed true. When self-pollinated, they give only the same kind of plant.
6 RR × rr gives all red plants (Rr), but these do not breed true because they have a hidden recessive (r) allele.
7 Rr × Rr gives RR, 2Rr, rr (3 : 1, red : white).
8 RR × Rr gives ½ RR : ½Rr. All appear red, but only half (RR) breed true, the other half (Rr) do not.
9 rr × Rr gives ½ rr : ½ Rr (½ white : ½ red)
10 Recessive alleles hidden in the F_1 may show again later.
11 Variation is continuous or discontinuous.
12 An important chromosome mutation is Down's syndrome.
13 Genetic engineering is the transferring of genes/alleles.

24 Selection: how organisms change

Aims of the chapter

After studying this chapter you should be able to:

K

1 Describe the effect of the environment on continuous and discontinuous variation.
2 Describe and give examples of natural selection.
3 Describe and give examples of artificial selection.

Variation and the environment

Continuous variation such as height and mass depend upon many genes. The development of such variation is most influenced by the environment. For example, a person may have the genes to grow tall, but if they have insufficient calcium for their bones their growth may be stunted.

Discontinuous variation such as tongue rolling and blood groups are determined by single genes. Such variation is least influenced by the environment. These characteristics are determined from fertilization and not subject to change.

The amount of influence the **environment** can have on any particular characteristic can be found out by comparing identical twins. These twins have identical genes, and so any differences between them must be due to the environment (both before and after birth). For example, they will both have the same blood group, but if one receives better nutrition and education they may score higher on an I.Q. (intelligence) test than the other.

In a similar way clones of identical plants can be grown in different conditions to find the effects of the environment.

Natural selection

One form of a gene may give the organism an advantage over the organisms with the other form of the gene. For example, in the peppered moth there are dark and light forms. Moths are eaten by birds. Light moths resting on lichen-covered trees are difficult to see and so fewer of them are eaten. But if pollution leads to loss of lichens and the trees become dark, then the light moths will be easily caught. Under these conditions the dark moths will be at an advantage and will increase in numbers. This is an example of

K

NATURAL SELECTION because it occurs in the natural surroundings.

In nature there is always competition between organisms, for food, space and mates. This means that **only the fittest will survive**, i.e. the ones best suited to the environmental conditions at the time. For example, antibiotics are given to kill bacteria. The bacteria are varied and there may be some which are not killed. These can reproduce and in time give rise to a strain of antibiotic-resistant bacteria.

Sickle cell anaemia is a disease of haemoglobin. People with the disease have joint pains, shortness of breath, fever and shorter life expectancy. They are homozygotes (SS). People with normal haemoglobin are homozygotes (AA). The heterozygotes (AS) do not suffer so much from the disease and they have some protection against malaria. So, although the sickle cell allele is a disadvantage in the homozygote (SS) it is an advantage in the heterozygote (AS) and so it stays in the population.

Artificial selection

K ARTIFICIAL SELECTION is when humans carry out selective breeding (choosing which organisms will produce offspring) so as to develop new varieties.

In any species there will be a lot of variation which has arisen in the past because of mutations. Humans select the organisms which show the characteristic they want to the greatest extent and breed these together. In the next generation they again select for the same characteristic and allow only those organisms to breed. Over a few generations the percentage of organisms with the desired characteristic increases. In the case of plants, asexual reproduction may then be able to maintain the selected characteristic. Some examples are:

In plants: increase in food production, disease-resistance, food value, early flowering, longer-lasting fruits.

In animals: increase in meat or milk production, appearance (as in various breeds of dogs).

Summary

1 Variation arises from mutations. Differences between organisms are due to genes and the environment.

2 Selection acts on variation. In nature there is 'survival of the fittest'. This leads to changes in the population.

3 Humans carry out artificial selection to produce new varieties of plants and animals suited to their needs.

Questions on Chapters 22–24

For questions **87–98**, each group of questions has five possible answers, (A), (B), (C), (D) or (E). For each question choose the best response. Each response may be used once, more than once or not at all.

Questions 87–90

(A) Meiosis (B) Mitosis (C) Gametes
(D) Fertilization (E) Body cells

87 A cell has 46 chromosomes. By what process does it produce cells with 23 chromosomes? ____
88 A cell has 23 chromosomes. After which process will it have 46 chromosomes? ____
89 Which kinds of cells are produced by meiosis? ____
90 Which process occurs in the human ovary? ____

Questions 91–94

(A) XX (B) XY (C) 22 pairs chromosomes plus XX
(D) 22 chromosomes plus X (E) 22 chromosomes plus Y

91 Which are the male sex chromosomes? ____
92 Which would be the chromosomes in an unfertilized egg? ____
93 Which would be the chromosomes in a girl baby? ____
94 Which would be the chromosomes in a sperm which would give rise to a boy baby? ____

Questions 95–98

(A) Heterozygous (B) Homozygous (C) Recessive
(D) Dominant (E) Phenotype

The following cross was carried out

Red flowered plant Rr	×	White flowered plant rr

With reference to the above cross:
95 The red flowered plant was ____
96 The description 'red flowered plant' is the ____
97 The red allele is ____
98 The white flowered plant was ____

99 Red and white flowered plants were crossed.

Red flowered plant	×	White flowered plant
Rr		rr

(a) (i) What gametes would be formed by the red flowered plant?
(ii) What gametes would be formed by the white flowered plant?
(b) What offspring would be formed from the cross? Show your working.
(c) If there were 100 offspring, how many of each colour would there be likely to be?

100 The following cross with fruit flies was carried out. Complete the diagram by adding the alleles present at each place. The allele for broad abdomen is B, and for narrow abdomen is b.

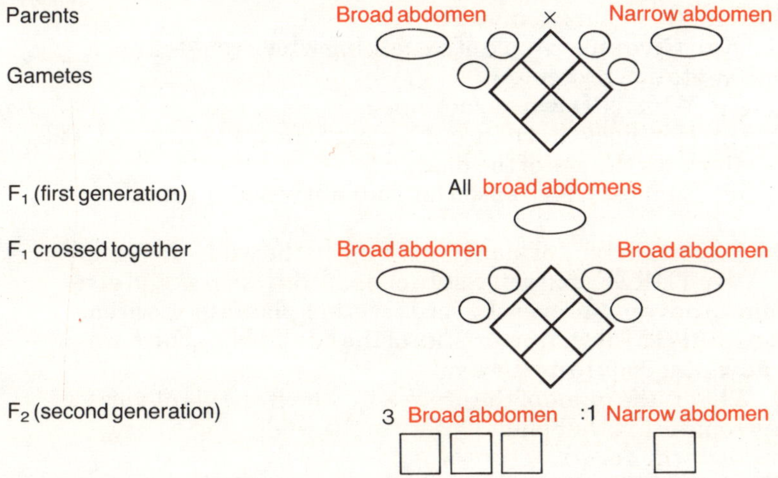

101 (a) In genetic engineering what part of the cell is transferred from one organism to another?
(b) List TWO problems which the experimenter might face.
(c) Give ONE characteristic in humans for which it would be an advantage to carry out genetic engineering.
(d) Give TWO reasons why it is difficult to carry out genetic engineering in humans.

102 In humans the brown-eye allele is dominant to the blue-eye allele. Explain how:
(a) Two blue-eyed parents cannot have a brown-eyed child.
(b) Two brown-eyed parents can have a blue-eyed child.

103 (a) (i) Give a definition of continuous variation.
(ii) Give a definition of discontinuous variation.
(b) From the following list of variations observed in the human population pick out TWO examples each of continuous and discontinuous variation.
(A) Height (B) Tongue rolling (C) Body mass
(D) Intelligence (E) Eye colour
(i) Examples of continuous variation ____ and ____
(ii) Examples of discontinuous variation ____ and ____
(c) Will continuous or discontinuous variation be most affected by changes in the environment?

104 (a) (i) What is a mutation?
(ii) Give an example of something which could cause a mutation to occur.
(b) (i) What is 'Down's syndrome'?
(ii) How does the chance of having a 'Down's syndrome' baby vary with the age of the mother?
(iii) Suggest ONE reason for your answer to (ii) above.

105 A certain kind of plant is growing in the wild.
(a) Give THREE characteristics of its seeds which would give them an advantage over the seeds of other plants in the area.
(b) Give THREE characteristics of the adult plant which would be advantageous to its survival.
(c) What is the name of the process by which one plant variety may replace another plant variety in the wild?

106 (a) (i) What is the name of the process by which humans have developed the various breeds of dogs?
(ii) Describe how a new breed of dog could be developed.
(b) List THREE reasons why it is easier to produce new varieties of plants than it is to produce new varieties of animals.

Now turn to page 154 to check your answers 87–106.

25 Assessment of practical skills

Aims of the chapter

After studying this chapter you should be able to:

1 Improve your practical skills of observing, measuring, drawing, calculating, presenting information, obtaining and interpreting information, carrying out experiments, and designing and evaluating investigations.
2 Identify areas of the syllabus tested in these ways.

Practical skills

Practical skills are tested by the teacher by the **Teacher Assessment** which counts for 20–30 per cent of the marks of the final examination. You will be assessed as you carry out your practical work. **But in addition most of the skills will also be tested on the written papers** so they will add even more to the final examination result. (Note that for external/mature candidates selected skills only will be tested on written papers. For more information refer to your particular syllabus.)

In the sections below the practical skills are described, and some of the ways in which they will be tested.

Observing

Observation involves using the senses to describe organisms and identify changes, similarities, differences etc. The observation is the major skill, and the recording is simple.
(a) **Describing organisms**, both whole organisms and parts of organisms (using a hand lens or microscope). You have to choose what to observe (both the overall features and the fine details), e.g. section of a leaf, wing of a fly.
(b) **Identify changes**, e.g. colour changes in food tests, volume changes, production of bubbles, changes in form, e.g. in life cycles, stages in germination.
(c) **Finding similarities** (matching). Listing similarities between organisms, or matching diagrams of organisms to descriptions (as when using a key to identify organisms).
(d) **Finding differences.** Listing differences between organisms, diagrams, photographs, e.g. comparison of flower structure of different species, comparing invertebrates.
K (NB. Your observations should be numbered. They might also be presented in the form of a table (see page 137).)

Measuring

K

Measurement skills involve the correct use of measuring instruments and reading of the scales.

(i) Thermometer. Allow 3 minutes for it to come to the correct reading. Read to the nearest 0.5° or 1°C.
(ii) Ruler or metre stick. Read to the nearest 1 mm (0.1 cm).
(iii) Measuring cylinder. On a flat surface, eye level with lowest part of curve (meniscus). Read to the nearest 1 cm^3.
(iv) Stopclock. Begin from zero mark. Read to nearest sec(s).
(v) Simple balance. Mass. Read to the nearest 1 g.

NB. When you record a reading make sure you also write down the units, i.e. °C, mm, cm^3, s or g.

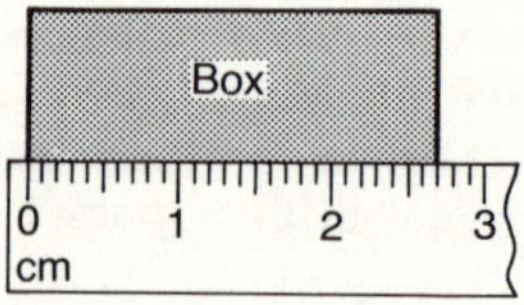

Length of box = 2.7 cm
or 27 mm

Drawing

K

Drawing is making a diagram, e.g. observing an organism and making a diagram to show the important parts which you then label. Here are some useful hints:

1 Use a sharp pencil and do not shade your drawing.

2 Identify the main parts and make a rough sketch first (this will help you get the proportions right).

3 Make it about twice the size of the object (i.e. ×2). So, if the object is 5 cm long make the drawing 10 cm long.

4 Rub out the outline as you fill in the details.

5 Draw straight label lines which do not cross.

6 Add the names and descriptions of the parts.

7 Give your drawing a title, and add the scale (e.g. ×2).

NB. You may also have to draw apparatus or a food web, etc.

Calculating

K

Calculations involve numbers. You will have to +, −, ×, and ÷ and work out averages, percentages and ratios.

Average (mean) $= \frac{\text{Total number}}{\text{Number of readings}}$ e.g. average of

$$3, 6, 9, 10, 5, 3 = \frac{3 + 6 + 9 + 10 + 5 + 3}{6} = \frac{36}{6} = 6$$

Percentage (%) = parts out of a hundred, e.g.
10.1 g of fibre in 100 g of cereal would be 10.1% fibre.
Similarly, 4.8% of protein would be 4.8 g in 100 g of food.

Ratios = relations between numbers, e.g. in genetics if there were 75 red plants and 25 white plants this is a ratio of 75 : 25 (which can be simplified to 3 : 1).
(NB. As there are 100 plants altogether, the percentage of red plants is 75% and of white plants is 25%.)

Presenting information

This covers writing notes and essays, and preparing flow charts, tables, bar charts, line graphs and piecharts.

(a) **Notes and essays** (see pages 157–8).

(b) **Flow charts.** Diagrams showing the order in which things should be done, e.g. testing a leaf for starch.

(c) **Tables.** The items to be described are listed down the left-hand side and the headings to the columns on the right-hand side. Then you add the information in the correct places.

Organisms	Similarity	Difference
X and Y		

(d) **Bar charts.** Along the upright (vertical) axis make a scale of the numbers (values). Along the horizontal axis enter the items to be described. Make bars of the correct heights.

Vegetables	% fibre
Potatoes	2.5
French beans	3.0
Sweetcorn	4.5
Peas	6.0

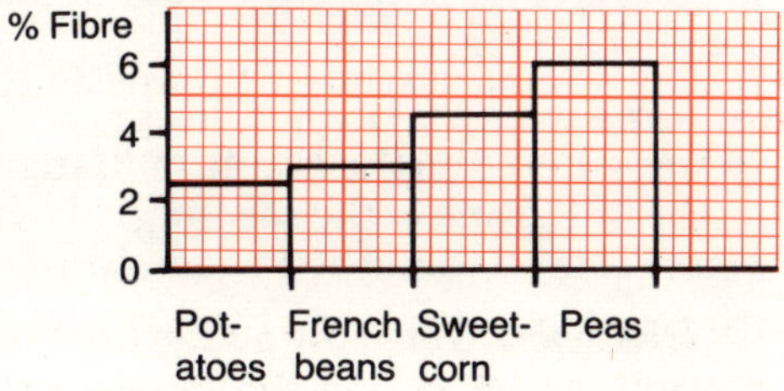

(e) **Line graphs.** Along the horizontal axis put the time (hours, days etc.) or other values you decide before the experiment. Up the vertical axis put the scale for the readings you take. Make a dot and circle to represent each reading and then join the dots by straight lines.

Seedlings	Mass (g)
Day 1	10
3	12
5	14
7	20

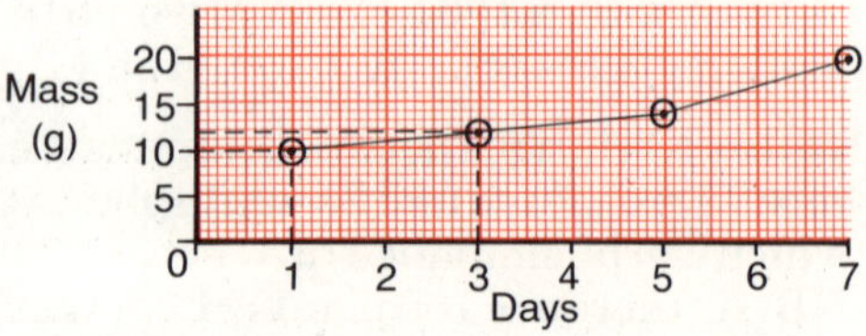

(f) **Piecharts** are like pies representing parts of the whole.

% constituent
Water 50%
Carbohydrate 25%
Protein 15%
Fat 10%

Angle at centre = $\frac{\text{\% constituent}}{100} \times 360°$

e.g. water sector = $\frac{50}{100} \times 360° = 180°$

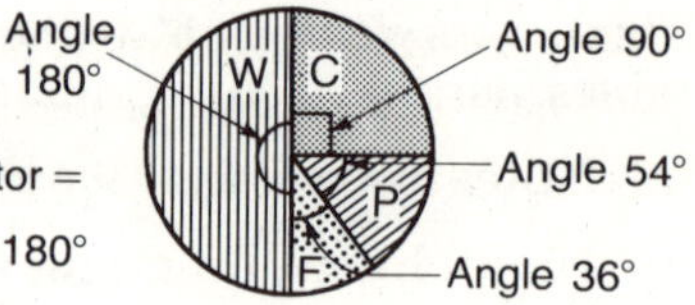

Obtaining and interpreting information

K

These skills are to test whether you understand written information, and can describe and draw conclusions from flow charts, tables, bar charts, line graphs and piecharts.

(a) **Written information.** Given a piece of scientific writing you would have to pick out the key points and explain information.

(b) **Flow charts.** Write out the procedures in sentences.

(c) **Tables,** e.g. identify particular readings from a table.

(d) **Bar charts.** Given a bar chart you would have to read off the heights of some of the bars. Your answer should include the units, e.g. % of fibre, g, etc.

(e) **Line graphs.** Given a line graph, you would be asked to read values entered on the graph, or to suggest values. For example, on the line graph of the seedlings above, you would guess that the mass on day 2 would be 11 g, and on day 6, 17 g.

(f) **Piechart.** Given the angle, e.g. 90°, you could work out the percentage. 90° is ¼ of 360°, so 90° represents 25%.

Carrying out experiments

This tests following instructions, carrying out procedures safely, and safe handling of equipment and living things.

(a) **Following instructions.** These can be written instructions or diagrams (e.g. flow charts) or verbal instructions.

(b) **Carrying out procedures safely.** You will be expected to do fieldwork, microscope work and general laboratory work, following any necessary safety precautions. You will be expected to show initiative, but you can ask for help.

(c) **Safe handling of equipment**, chemicals and living things. You will be expected to set up, to use and also to describe the use of certain items. For example:

(i) **Use of the microscope**
1 Adjust the mirror and iris diaphragm so that there is a clear background when viewed through the eyepiece.
2 Clip the glass slide with the specimen onto the stage.
3 *Looking from the side,* lower the body tube with the coarse adjustment so the objective lens is close to the slide.
4 *Looking through the eyepiece,* slowly raise the body tube with the fine adjustment until the specimen comes into focus.

(ii) **Heating solutions**
1 Clear items away from the area, and make sure your hair and clothing are not too near the Bunsen flame.
2 Wear safety goggles if necessary.
3 Hold the test tube in tongs (not in your fingers) because the tube will get hot. Move it around in the flame.
4 Hold the mouth of the test tube so that it points away from yourself and away from other people.
5 After heating, rest the test tube in a test tube rack.

(NB. Flammable liquids such as ethanol must not be heated with a flame. Put the test tube with ethanol in a beaker of boiling water (with the Bunsen turned off).)

Designing and evaluating investigations

You have to describe how investigations should be set up (with controls), criticize experiments, and give conclusions that can be drawn. Most questions involve correct use of controls and of making the tests 'fair'.

(a) **Designing (planning) investigations.** You may be given a hypothesis (untested theory) to investigate, and you have to set up an experiment to test it. Some hints:

1 Use large numbers, e.g. of seeds or fruits so that any differences are not likely to be due just to chance.

2 List the apparatus you will need. Keep things simple.

3 Set up alternative experiments in identical ways, except for the conditions (variables or factors) to be examined, e.g. the same numbers, same mass, same volume, same amount of light, same temperature, same number of readings etc.

4 Decide ahead of time how you will judge a 'good' result, e.g. when germinated, when clean, when mouldy etc.

K The CONTROL has all the conditions considered necessary. Each experiment must differ in just one way from the control.

(b) **Criticizing experiments**

1 You may be asked to account for a certain procedure, such as use of sterile test tubes, black cover on a plant.

2 You may be asked what difficulties may be expected.

3 You may be asked to describe the control to be used.

4 Criticisms may also be based on incorrect controls, so that the test is not a fair one. You will be asked to suggest what changes would be needed to make the experiment fair (see (a) above).

(c) **Drawing conclusions.** You may be given information (data) in written form or as tables, line graphs, bar charts etc. You will have to give reasons for the results and conclusions. Note:

1 The conclusions you can correctly draw will depend upon how well the original experiment was set up. It should have had a control, so that any differences observed will be true differences and not just due to chance.

2 Some of the explanations of the experiments will depend upon your having done the experiment yourself.

Summary

1 Practical skills are tested during practical work and also in the written examinations.

2 Skills covered include observing; measuring; drawing; calculating; presenting and interpreting information in written form, flow charts, tables, bar charts, line graphs and piecharts; carrying out experiments and designing and evaluating investigations.

Answers to questions

Note. Don't look at these answers until *after* you have done the questions. If you peep at them beforehand, the only person you are cheating is yourself.

With each answer there are some 'Hints' which should help you in answering similar questions.

On the right-hand side of the page are the number of marks for each part of the question. Use the answers to find your own score. Give yourself a red tick for each mark you get. Then at the end add up all the ticks. If you think you have a correct answer, but it is not listed then check it with your teacher.

Answers to Questions 1–7 (pages 18–19)

1 (i) **R**espiration: release of energy from food breakdown. (2)
(ii) **E**xcretion: removal of waste products of metabolism. (2)
(iii) **M**ovement: getting from one place to another, or movement of cell contents. (2)
(iv) **I**rritability: sensing changes and responding to them. (2)
(v) **N**utrition: getting food (making it or eating it). (2)
(vi) **D**evelopment or growth: increase in size/complexity. (2)
(vii) **R**eproduction: making new organisms similar to parents. (2)

Hints: The characteristics can be given in any order.

2 A (1)
3 C (1)
4 E (1)
5 B (1)

Hints: You need to know the characteristics of all the major animal and plant groups.

6

(a)		(S)	(T)	
Differences	(i)	3 pairs of legs	5 pairs of legs	(1)
	(ii)	Wings	No wings	(1)
	(iii)	No pincers/claws	Pincers/claws	(1)
	(iv)	Separate parts	Parts not separated	(1)

Hints: You are not being asked to recognize a cockroach and a crab, you just have to observe the differences.

(b)		(R)	(U)	
Differences	(i)	No legs	Legs	(1)
	(ii)	No antennae	Antennae	(1)
	(iii)	Clitellum/saddle	No clitellum/saddle	(1)

(c) (i) (S), (T) and (U). Hint: The most similar organisms (1)
(ii) I Legs/jointed legs (1)
II Antennae/feelers (1)

7 (A) *Quercus* (B) *Syringa* (C) *Fraxinus* (1 each
(D) *Platanus* (E) *Alnus* (F) *Aesculus* = 6)

Hints: Go through the key a little at a time. The first step is into 'Leaf divided into separate leaflets' (this is C and F), and 'Leaf not divided into leaflets' (this is A, B, D and E).

Write the letters on the key near to the descriptions.

Then go on to the next step. 'Five leaflets', is either C or F. It is F. So fill in the name 'Aesculus' near to F in the answers. And now you see that 'Fraxinus' is C. And so on.

Do not be put off by the Latin names, they are used so that you have to use the key and not just guess the answers.

For Questions **1–7** there are 34 marks. How well did you do?		
14 marks and less	Poor	Learn the topics again and do all the questions a second time.
15–19 marks	Fair	
20–23 marks	Average	Check the questions you got wrong and do them again.
24–27 marks	Good	
28–31 marks	Very good	Check through your mistakes.
32 marks and over	Excellent	

Now go back to page 20 for the next part of the text, or to page 48 to do the next lot of questions.

Answers to Questions 8–31 (pages 48–50)

8	D	Hint: You need to know these definitions.	(1)
9	C		(1)
10	A		(1)
11	B		(1)
12	C		(1)
13	E		(1)
14	B	Hint: 'Density' means 'number per square metre'.	(1)
15	B	Hint: Note that alternative 'B' is used twice.	(1)
16	E	Hint: You have to understand about trophic levels. If you got them wrong check page 28.	(1)
17	A		(1)
18	D		(1)
19	B		(1)

20 C or E Hint: either answer is correct. (1)
21 A Hint: However, some plastics are biodegradable. (1)
22 E or D (1)
23 E Hint: Not C, as it may carry infections. (1)
24 (a) Plant leaves→Aphids→Ladybirds→Sparrow→Hawks,
or, Leaf litter→Millipedes→Centipedes→Sparrows→Hawks } (1)
(b) **Either** Aphids **or** Slugs **or** Earthworms **or** Millipedes (1)
Hint: Primary consumers eat producers (the plants).
(c) (i) The aphid population would increase. (1)
After some time it may decrease due to lack of food (1)
(ii) Sparrows have less food, so numbers decrease. (1)
Or eat more slugs, earthworms and centipedes. (1)
Hints: Look at other food webs and make yourself questions of the same kind. Use the ideas in the answers above.

25 B Hint: If you got this wrong, look at page 28. (1)

26 A Hint: At each trophic level, energy is lost. So the shortest food chain has least energy loss. (1)

27 (a) Sulphur dioxide (1)
Kills plants, for example lichens, **or**
Damages respiratory system of animals, **or**
Eats away at buildings } (1)
(b) Excess nitrates run into freshwater (1)
Nitrates in water cause increase in algae numbers (1)
Algae decomposed by bacteria (1)
Bacteria use up oxygen and fish die. (1)
Hints: Make sure you know this connection between excess nitrates and dead fish. Excess nitrates also from sewage.

28 Following or similar points:
(Any THREE, 1 mark each, total 3)
Clearing land for housing removes animals' habitats.
Manure in water takes up oxygen and run-off kills fish.
Polluting air with acid rain kills plants.
Wastes dumped in rivers damage aquatic life.
Using pesticides kills other animals as well.
Sprays and aerosols pollute atmosphere, affect animals.
Fishermen leave lead in water, kill swans etc..
Oil pollution affects seabirds.
Farmers using herbicides, kill other organisms.
Acid rain makes soil acid and bacteria less active.
Burning ground, harms soil and soil organisms. (3)

29 (a) Process C Absorption **or** Uptake (1)
Process D Nutrition **or** Feeding **or** Eaten by (1)

(b) (i)	B and F	Hints: For this question you don't have	(2)
(ii)	B	to know the names of bacteria. You	(1)
(iii)	A	just have to understand the diagram.	(1)

(c) E Hint: This is the 'decay' stage also carried out by saprophytic fungi by external digestion. (1)

Hints: You will also need to revise the carbon cycle and the water cycle. Revise them on pages 31 and 34 and then set yourself questions on the various processes involved.

30 (a) Following or similar points:
(Any FOUR, 1 mark each, total 4)
Food supply or prey
Enemies or predators
Disease if overcrowded
Competition for space or mates
Accumulation of waste products (may be poisonous)
Amount of rainfall (e.g. if a drought) (4)

(b) **Either** by war **or** murder **or** birth control (1)

(c) **Either** by use of medicines (or useful drugs) **or** hospitals **or** fertility drugs **or** social services (1)

Hints: Parts (b) and (c) are to find out if you can think of reasonable answers.

31 (a) Parasite: An organism which **feeds** on another, **living** organism. (2)

(b) Antibiotic: A **drug** given to **kill bacteria** which are causing disease. (2)

(c) Antibodies: produced by some **white** cells to **kill disease** organisms. (2)

For Questions **8–31** there are 52 marks. How well did you do?

14 marks and less	Poor	Learn the topics again and do all the questions a second time.
15–18 marks	Fair	
19–26 marks	Average	Check the questions you got wrong and do them again.
27–34 marks	Good	
35–42 marks	Very good	Check through your mistakes.
43 marks and over	Excellent	

Now go back to page 51 for the next part of the text, or to page 71 to do the next lot of questions.

Answers to Questions 32–52 (pages 71–4)

32 D (1)
33 B Hint: If the plant didn't have a cell wall it couldn't (1)
34 B become turgid. (1)
35 C (1)
36 A (1)
37 B (1)
38 D Hint: Anaemia is caused by poor haemoglobin. (1)
39 C (1)

40 (a) **Same** as diagram, but **boil** amylase first. (2)
(b) (i) Boil with Benedict's **or** Fehling's solutions. (1)
(ii) The water in the beaker should be tested. (1)
(iii) Starch has been changed to a simple sugar. (1)
(c) (i) Similar. Any 2 of: Selectively (semi-) permeable, Digestion of starch occurs, Contains amylase. (2)
(ii) Different. Any 2 of: Non-living, No blood supply, Does not have villi. (2)
Hints: This apparatus is sometimes called a 'model gut'.

41 (a) Enzyme (1)
(b) D or 30°C (1)
(c) More activity with higher temperature. (1)
But at 50°C its activity would be destroyed. (1)

42 (a) Any 5 of the following or similar points:
1 Not used up **or** Can be used again
2 Speeds up **or** Changes rate of reaction
3 Made of protein
4 Affected by temperature **or** Destroyed by heating
5 Affected by pH **or** by acids or alkalis
6 Specific to a particular kind of reaction
7 Affected by poisons (5)

(b) e.g. Temperature. **Compare rate** of activity at **3** different temperatures, e.g. 0°C, room, boiled **(control)**. (3)
Hint: In any answer include compare rate, 3 variations, control.

43 (a) Starch with Iodine solution goes blue-black (2)
(b) Oxygen Relights a glowing splint (2)
(c) Carbon dioxide Turns limewater milky (2)
(d) Glucose Heat with Benedict's or Fehling's solutions, red precipitate formed (2)

44 C Hint: Oxygen is released, not used. (1)
45 B Hint: Carbon dioxide was limiting photosynthesis. (1)
46 (a) and (b) see diagrams below.

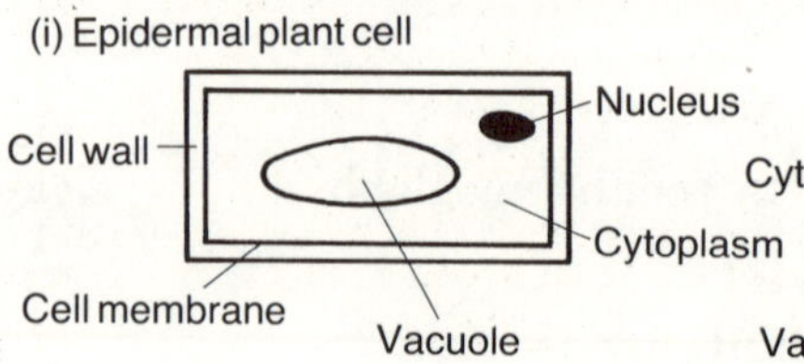

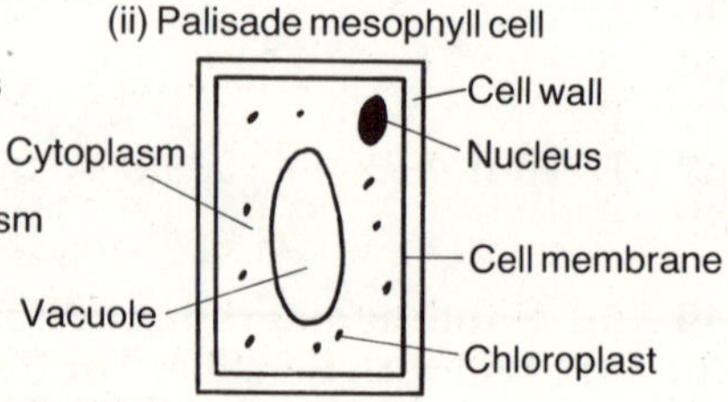

½ mark for each label, to a total of 5 marks (5)

(c)	Typical animal cell	Palisade mesophyll cell	
(i)	Irregular shape	Definite shape	
(ii)	No cell wall	Cell wall	
(iii)	No chloroplasts	Chloroplasts	
or	Small vacuoles	Large vacuoles	(3)

47 (a) 28.6 g Hint: 28.6% means 28.6 parts out of 100. (1)
(b) 11.44 g
Hint: In 100 g of cereal there are 28.6 g fibre so in 40 g there will be $28.6/100 \times 40 = 11.44$ g (3)
(c) Fibre prevents constipation, **or**
Encourages movement of food in the gut (1)

48 (a) (i) C (1)
(ii) Highest vitamin C content (1)
Not much protein, fat **or** carbohydrate, so it isn't something that would supply much energy (1)
(b) (i) B (1)
(ii) Highest in protein needed for growth of baby (1)
Quite high in calcium needed for bones and teeth (1)
Highest in iron needed for making haemoglobin (1)
(c) (i) D (1)
(ii) D has the highest % of fat, which gives much more energy than proteins or carbohydrates (1)
Hint: You don't need to do the arithmetic, you just have to think about the problem to get the answer.
(d) (i) Scurvy (1)
(ii) Bleeding gums **or** internal bleeding (1)

49 *Parts* *Functions*

(a) Incisors	Cutting off food or Biting food		(1)
(b) Molars	Chewing or Grinding food		(1)
(c) Saliva	Begins starch digestion or Alkaline to neutralize acids or Food easy to swallow		(1)
(d) Oesophagus	Peristalsis or Passes food to stomach		(1)
(e) Stomach	Enzymes or Acid softens food		(1)
(f) Liver	Produces bile	Hint: Related to digestion	(1)
(g) Pancreas	Produces enzymes		(1)
(h) Sm. intestine	Absorbs food or Contains villi		(1)
(i) Colon	Absorbs water		(1)
(j) Rectum	Stores faeces before egestion		(1)

50 (a) Breakdown into **smaller** pieces of the **same** chemical nature (2)
e.g. Action of teeth **or** Action of bile on fats (1)
(b) Breakdown into **small** parts of a **different** chemical nature **or** breakdown by enzymes (2)
e.g. Digestion by enzymes **or** Making particles soluble (1)

51 (a) (i) Brand A (1)
(ii) 3 of the following: Less fat, Less carbohydrate Artificial sweetener, Less energy (3)
(b) (i) Milk (1)
(ii) In a refrigerator (1)
(iii) No. Growth is slowed down in the fridge, but bacteria will grow and eventually make yoghurt go bad (2)
(c) (i) Protein 6 g. In 100 g there are 4.8 g protein.
In the additional 25 g (a ¼ of the whole) there are
4.8/4 = 1.2 g. So, 4.8g + 1.2 g = 6 g. (2)
(ii) Fat 0.25 g. In 100 g there are 0.2 g fat.
In the additional 25 g there will be 0.2/4 = 0.05 g
So, 0.2 g + 0.05 g = 0.25 g (2)

52 (a)

Enzyme groups	Substrate	End-product	
Proteases	Proteins	Amino acids	(2)
Lipases	Fats or Lipids	Fatty acids and glycerol	(2)
Amylases	Carbohydrates	Glucose	(2)

(b) (i) Graph C (1)
(ii) Acid conditions = low pH (1)
Graph C shows most activity at low pH (1)

For Questions **32–52** There are 100 marks. How well did you do?		
19 marks and less	Poor	Learn the topics again and do all the questions a second time.
20–26 marks	Fair	
27–39 marks	Average	Check the questions you got wrong and do them again.
40–52 marks	Good	
53–65 marks	Very good	Check through your mistakes.
66 marks and over	Excellent	

Now go back to page 75 for the next part of the text, or to page 91 to do the next lot of questions.

Answers to Questions 53–69 (pages 91–4)

53 (a) (i) D **or** Left ventricle (1)
Extra power needed to send blood around the body (1)
(ii) A and C **or** Right atrium and Right ventricle (2)
Hint: **R**ight side of heart – **D**e-oxygenated blood
Left side of heart – **O**xygenated blood
(Memory aid: **R**eader's **D**igest/**L**eft **O**vers)

(b) I pulmonary artery (1)
II pulmonary vein (1)
III vena cava (1)
IV aorta (1)
Hint: Make sure you can also say which chamber of the heart they are associated with.
(ii) II and IV Hint: Associated with left side of heart (2)
(iii) Statements 3 and 4 are correct (2)

(c) (i) Coronary arteries (1)
(ii) Two of: Lack of oxygen or food or blood, **or** Angina **or** Heart attack (2)

54 A (1)
55 E (1)
56 B (1)
57 E (1)
Hint: If you got any of these wrong check the functions of the blood, on page 78.

58 (a) D **or** xylem (b) C **or** phloem (2)
(c) (i) Photosynthesis (ii) Translocation (2)
(iii) Absorption **or** Osmosis (1)
(iv) Transpiration **or** Evaporation **or** Dehydration (1)
(d) (i) Increase in temperature. Wind. Dry air. (3)
(ii) Warm air can contain more water vapour, **or**
Moving air removes saturated air, **or**
Dry air can contain more water vapour (1)
(e) Plant wilts **or** Becomes dehydrated **or** Stomata close (1)

59 B } Hint: These are sometimes confused with (1)
60 A } each other. (1)
61 E Hint: Urea is *made* in liver, *excreted* from kidney. (1)
62 D (1)
63 F (1)
64 E (1)

65 (a) A (1)
(b) B and C Hint: No oxygen in equation. (2)
(c) (i) In animal muscle (*not* just in animal cells) (1)
(ii) Not enough oxygen **or** Because of exercise (1)
(iii) Fatigue in muscle **or** Oxygen debt or deficiency **or** Panting **or** Build-up of lactic acid (1)
(d) (i) Yeast **or** Germinating seeds **or** Fermentation (1)
(ii) Yeast for fermentation **or** bread making **or** alcohol (1)

66 (a) (i) Thin (ii) Moist (iii) Large surface (3)
(b) (i) Thin cell walls – short distance for diffusion (1)
(ii) Moist – gases can dissolve (1)
(iii) Large surface – large area for gas exchange (1)
(c) (i) Supplied with blood, and Breathing movements (2)
(ii) Any 2: Gas exchange quicker, Animal larger, Animal more active, Diffusion too slow, Needs more oxygen (2)

67 (a) (i) Inhaled air (ii) Control (2)
(b) (i) Limewater (1)
(ii) It goes cloudy (1)
(iii) CO_2 in exhaled air **or** CO_2 reacts with limewater (1)

68 (a) (i) Heart **or** pulmonary artery (ii) Oxygen (2)
(iii) Carbon dioxide (iv) Heart **or** pulmonary vein (2)
(b) Any 3: Bronchitis, Poor breathing, Increase in heart disease, Lung cancer, Yellow teeth, Throat cancer (3)

69 (a) I Kidney – removal of urea **or** water balance (2)
II Ureter – takes urine to bladder (2)
III Bladder – stores urine (2)
IV Urethra or penis – passes out urine (urination) (2)

(b)

	A	B	
(i)	More oxygen **or** food	Less oxygen **or** food	(1)
(ii)	More urea **or** salts	Less urea **or** salts	(1)
(iii)	Less carbon dioxide	More carbon dioxide	(1)

For Questions **53–69** there are 77 marks. How well did you do?		
28 marks and less	Poor	Learn the topics again and do all the questions a second time.
29–38 marks	Fair	
39–46 marks	Average	Check the questions you got wrong and do them again.
47–54 marks	Good	
55–62 marks	Very good	Check through your mistakes.
63 marks and over	Excellent	

Now go back to page 95 for the next part of the text, or to page 108 to do the next lot of questions.

Answers to Questions 70–78 (pages 108–10)

70 (a) Grows towards light **or** Bends towards light (1)
(b) As experiment, but with top covered with something light-proof **or** Light from all directions (2)
(c) (i) Similar: Useful for survival. Response to stimulus (2)
(ii) Different: No nerves, Slower, Control by auxin (2)

71 (a) (i)–IV, (ii)–I, (iii)–II, (iv)–III (4)
(b) (A) sensory nerve fibre or neurone (B) spinal nerve (C) spinal cord (D) intermediate or relay or connecting neurone (E) motor nerve fibre or neurone 1 mark each (5)
(c) Avoid danger or Protect from injury (1)

(d)	Spinal reflex	Voluntary action	
Differences	Quicker	Slower	(1)
	Without thought	Involves thinking	(1)

72 (a) (A) Iris (B) Pupil (C) Cornea (D) Lens (E) Ligament (F) Ciliary muscles (G) Vitreous humour (7)
(b) (i) Becomes larger **or** wider in dim light (1)
(ii) Any 2: Change in shape of iris, Reflex action, Circular muscles relax, Radial muscles contract (2)
(c) (i) (E) **or** Suspensory ligaments slacken as (1)
(F) **or** ciliary muscles contract (1)
(ii) To focus close objects **or** To read (1)

(d) (i)			
I	Sclera	Outer protective layer	(2)
II	Choroid	Has blood vessels **or** Brings nourishment	(2)
III	Retina	Sensitive to light **or** Has rods and cones	(2)
IV	Yellow spot	Highest concentration of cones	(2)
V	Blind spot	Exit place of optic nerve	(2)

(ii) Impulses from retina to brain (1)
Brain interprets image as 'right-way up' (1)

73 B (1)
74 C (1)
75 E (1)
76 A (1)

77 (a) (A) Bone **or** Humerus
(B) Muscle **or** Triceps
(C) Muscle **or** Biceps
(D) Tendon
(E) Ligament
(F) Cartilage
(G) Fluid **or** Synovial fluid
1 mark each (7)

(b) (E) Hold the joint together (1)
(F) Soft and absorbs shock (1)
(G) Lubricates joint **or** Allows easy movement (1)

(c) (B) Relaxes **or** Stretches **or** Extends (1)
(C) Contracts **or** Shortens (1)
(D) Nothing **or** Inelastic **or** Pulls on bone (1)

(d) Any 3: Protection, Support, Production of red cells, Framework, Stores calcium (3)

78 (a) (A) Epidermis
(B) Dermis
(C) Fat **or** fat layer
(D) Nerve ending
(E) Blood vessel **or** capillary
(F) Sebaceous gland
(G) Hair
(H) Muscle **or** erector muscle
(I) Blood vessel **or** capillary
(J) Sweat gland (10)

(b) (E) Widen, so more blood comes to skin to be cooled (2)
(H) Relaxes, so hairs flat and more heat is lost (2)
(I) Widen, so more liquid to sweat gland (2)
(J) More sweat, evaporation cools body (2)

(c) (i) D Hint: Sensitive to different stimuli (1)
(ii) C (1)
(iii) H **or** G Hint: Hairs raised to keep warm (1)
(iv) J (1)

For Questions **70–78** there are 86 marks. How well did you do?

24 marks and less	Poor	Learn the topics again and do all the questions a second time.
25–33 marks	Fair	
34–39 marks	Average	Check the questions you got wrong and do them again.
40–45 marks	Good	
46–52 marks	Very good	Check through your mistakes.
53 marks and over	Excellent	

Now go back to page 111 for the next part of the text, or to page 121 to do the next lot of questions.

Answers to Questions 79–86 (pages 121–3)

79 (a) Any 3: Increase in mass **or** dry mass.
Increase in length of root **or** height of stem.
Increase in number **or** size of leaves.
Increase in width of stem. Any 3, 1 mark each (3)

(b)

Plant growth	*Animal growth*	
Continuous through life	Stops when mature	
Branching shape	Compact shape	Any 4 points
Occurs at apices	Occurs all over body	(4)

80 (a)

Asexual reproduction		*Sexual reproduction*		
Bulbs	Runners	Stamens	Petals	(½ × 8
Corms	Spores	Ovules	Gametes	= 4)

(b) Can get identical plants **or** same variety (1)

81 (a) Any 3: Reduced **or** dull **or** green **or** absent petals.
No scent **or** No nectar. Smaller **or** lighter pollen.
Filaments long **or** anthers hang out of flower.
Long **or** feathery stigma. Any 3, 1 mark each (3)
Hint: You should also know the reasons for the differences.

(b) (i) Put immature (undeveloped) flower (1)
In a bag **or** On its own (1)
(ii) From immature (undeveloped) flower (1)
Remove the stamens (1)

82 (a)

(A)	Penis	Used during sexual intercourse	(2)
(B)	Urethra	Passes sperm **or** passes urine	(2)
(C)	Ureter	Urine from kidney to bladder	(2)
(D)	Bladder	Stores urine	(2)
(E)	Sperm duct	Transports sperm	(2)
(F)	Prostrate gland	Secretion activates sperm	(2)
(G)	Seminal vesicle	Secretion supplies nutrients	(2)
(H)	Scrotum	Protects testes **or** Keeps them cool	(2)
(I)	Testis	Produces sperm **or** Male sex hormone	(2)

(b) Any 3: (D), (B), (C) and (A) 1 mark each (3)

(c) (i) Sperm cannot now pass from testis to urethra, **or**
Male becomes infertile (1)
(ii) Some sperm still alive (1)
In sperm duct **or** seminal vesicles (1)

Hint: Also 2 marks for: Man may have donated sperm before the operation which were later used for fertilization.

83 (a)
(A) Uterus **or** womb Development of embryo (2)
(B) Oviduct **or** Fallopian tube Path for egg **or** Fertilization (2)
(C) Ovary Produces eggs (2)
(D) Cervix Opening of uterus (2)
(E) Vagina Passage to outside **or** Sperm deposited (2)
(F) Vulva Opening of vagina (2)
(b) Any 2: Urethra needed just to excrete urine.
Vagina has to receive penis **or** for sexual intercourse.
Vagina has to stretch for birth of baby. (2)
(c) (i) Lining shed **or** Cast off **or** Bleeds (1)
(ii) Lining built up **or** Repaired **or** Receives more blood (1)
(iii) Remains supplied with blood **or** Is not cast off
or Forms the placenta **or** Remains thick (1)
(d) Any 4: Insufficient sperm. Damaged sperm. S.T.D.s.
Eggs not produced. Blocked Fallopian tubes.
Sterilization. Menopause. (4)

84 (a) Arrow away from foetus in umbilical artery (1)
Arrow towards foetus in umbilical vein (1)
Hint: **A**rteries **A**way from heart
(b) (i) Any 2: Carbon dioxide. Wastes. Urea. Water. Heat. (2)
Hint: Substances and things foetus is getting rid of.
(ii) Diffusion. Hint: Substances tend to move along their diffusion gradients. (1)
(c) (i) Any 2: Food, Oxygen, Antibodies. (2)
Hint: Substances passing from mother to foetus.
(ii) Any 2: Drugs. Alcohol. Nicotine. Viruses. (2)
(d) (i) Cord is tied in two places and cut, **or**
Drops off the baby in due course (1)
(ii) Placenta passes out of the body, **or**
Passed out as the 'afterbirth' (1)

85 (a) (A) Remains of style and stigma (1)
(B) Ovary wall (1)
(C) Ovule or fertilized ovule (1)
(E) Remains of sepals (1)
(b) (C) has been fertilized and grown into a seed, but (1)
(D) has not been fertilized, and not grown (1)
Hint: NOT (C) is older than (D), nor (D) is only partly developed.
REASON is required.
(c) By explosive mechanism, **or** As the fruit wall dries
and breaks, seeds will be shot out (1)

86 (a) (i) Warmth Hint: 22°C (1)
(ii) Moisture **or** Water **or** Damp (1)
(iii) Darkness **or** No light (1)
(iv) Air **or** Oxygen Hint: Air space in test tube (1)
(b) Any 2: One seed might fail to grow. Fairer.
To compare the seeds. Greater chance of success. (2)

For Questions **79–86** there are 88 marks. How well did you do?

26 marks and less	Poor	Learn the topics again and do all the questions a second time.
27–35 marks	Fair	
36–42 marks	Average	Check the questions you got wrong and do them again.
43–49 marks	Good	
50–56 marks	Very good	Check through your mistakes.
57 marks and over	Excellent	

Now go back to page 124 for the next part of the text, or to page 132 to do the next lot of questions.

Answers to Questions 87–106 (pages 132–4)

87 A Hint: Meiosis is reduction division. (1)
88 D (1)
89 C Hint: Meiosis only occurs in gamete formation. (1)
90 A Hint: In production of gametes, the eggs (1)
91 B (1)
92 D Hint: Not E. **Egg** has to be from female. (1)
93 C (1)
94 E (1)
95 A (1)
96 E (1)
97 D (1)
98 B (1)
(95–98) Hint: You should know these definitions.

99 (a) (i) R and r (ii) r and r **or** All r (2)
(b)

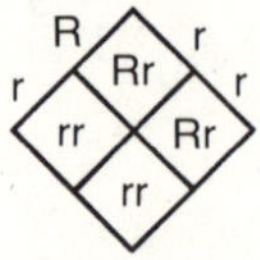

or

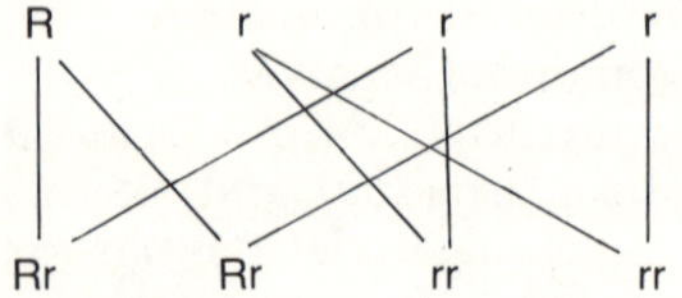

(3)

Equal numbers of red and white flowered plants (1)
(c) About 50 red and 50 white (1)

100 Parents — Broad abdomen × Narrow abdomen (2)

Gametes (3)

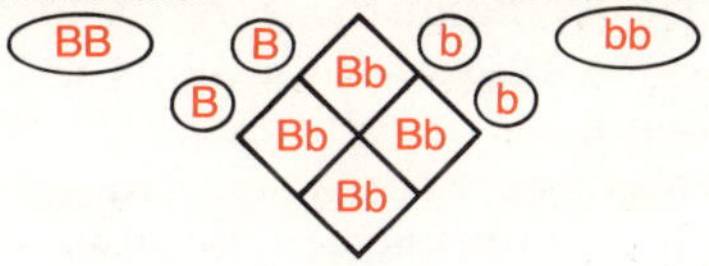

F_1 (first generation) — All broad abdomens

Bb (1)

F_1 crossed together — Broad abdomen — Broad abdomen (2)

Gametes (3)

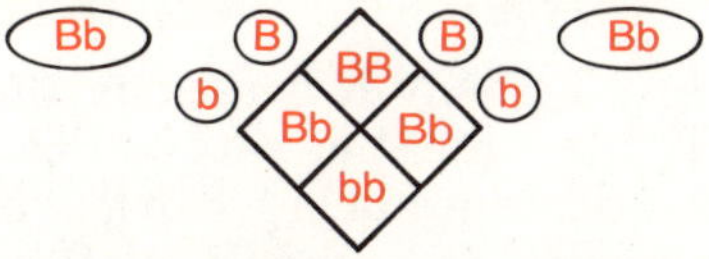

F_2 (second generation) — 3 broad abdomens : 1 Narrow abdomen

BB Bb Bb — bb (4)

101 (a) Genes **or** alleles **or** part of chromosomes (1)
Hint: No marks for 'nucleus'.
(b) Any 2: Cells **or** chromosomes **or** genes very small.
Difficult to identify the part to transfer.
Have to do the transfer without killing the cells.
May create something dangerous. (2)
(c) Any genetic disease, for example
Haemophilia **or** Sickle cell anaemia **or** Diabetes (1)
(d) Any 2: Right genes are hard to find.
Difficult to introduce genes into humans.
Complex human system or body.
Don't want to tamper with human development.
Danger of causing harm to humans. (2)

102 (a) Blue-eyed allele is recessive, so blue-eyed parents only have blue alleles and so they can't have a brown-eyed child (2)
(b) Brown-eyed parents may not be homozygous for brown alleles. Blue-eye alleles may be hidden, which can come together in a child, which will have blue eyes. (2)

103 (a) (i) Gradual changes in population from one extreme to the other **or** There are always intermediates (2)
(ii) Distinct groups **or** With or without a characteristic **or** No 'in-betweens' **or** No intermediates (2)
(b) (i) Continuous (A) and (C) (Also (D)) (2)
(ii) Discontinuous (B) and (E) (2)
(c) Continuous Hint: Under control of many genes (1)

104 (a) (i) A change in a gene **or** chromosome (1)
(ii) X-rays **or** Radiation **or** Chemicals **or** Nuclear bomb (1)
(b) (i) An extra chromosome Hint: This used to be called Mongolism, but this term is not now used. (1)
(ii) Chance of having a child with Down's syndrome increases with the age of the mother. (1)
(iii) More likely egg production is faulty with age (1)

105 (a) Any 3: Easily dispersed **or** A method of dispersal. Quick growing. High germination rate **or** Faster germination. Resistance to disease. Lot of green leaves. (3)
(b) Any 3: Early flower production. Producing many seeds **or** fruits. Growth of deep roots. Resistance to disease. Attractive flowers. (3)
(c) Natural Selection **or** 'Survival of the fittest' **or** Evolution (1)

106 (a) (i) Artificial selection **or** Un-natural selection (1)
(ii) From the population choose dogs with the characteristics similar to those you want. (1)
Breed these together. (1)
Choose from them the ones you want for the next generation or those with the characteristics most well developed. (1)
Breed from these again. Keep on in this way. (1)
(b) Any 3: Plants usually produce more offspring. Easier to cross the ones required. Usually have a shorter life cycle. Usually can be self-pollinated. Can reproduce asexually to keep the variety pure. (3)

For Questions **87–106** there are 73 marks. How well did you do?

12 marks and less	Poor	Learn the topics again and do all the questions a second time.
13–17 marks	Fair	
18–26 marks	Average	Check the questions you got wrong and do them again.
27–35 marks	Good	
36–45 marks	Very good	Check through your mistakes.
46 marks and over	Excellent	

DON'T FORGET! GO BACK OVER AND CHECK YOUR MISTAKES.

The week before

1 Plan out some essay questions in spider form (below.) Then use the diagram to write an essay.

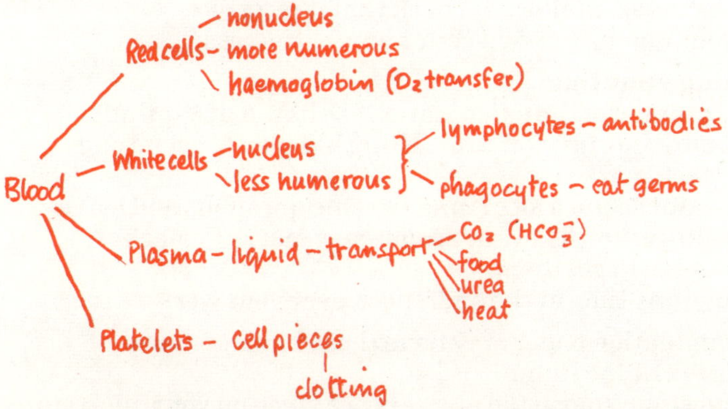

2 Go through this book. For every diagram in turn, cover over the labels and then try to put the correct labels back.
3 Try to write a function for every label on the diagram, or describe the importance of the structure or process.
4 Re-do the questions in this book and carefully check the answers. Take special care with the ones you get wrong.
5 Try to make your own memory aids, like REMINDeR (for life processes), and Reader's Digest (right side heart: de-oxygenated blood), Left Overs (left side heart: oxygenated blood).
6 Re-check the chapter on Assessment of Practical Skills, especially the parts on bar charts, line graphs and controls.
7 Look at specimen questions or past papers from your examination group and answer the questions.

The night before

1 It is too late to read textbooks. You need quick summaries.
(a) Skim through the revision book you have made, looking at your own notes and diagrams.
(b) Skim through the questions and answers in this book.
2 Write your exam number on paper or on your hand.
3 Collect together your pen, pencils, eraser, pencil sharpener, ruler, calculator etc. and make sure they are ready to use.
4 **Don't** stay up late. It will make you muddled next day.

The examination day

1 Get to the examination room in good time; avoid a rush.
2 Don't be put off by the 'exam nerves' of other students.
3 In the exam room make sure you have the correct paper.
4 Add your index number, or check that it is correct.
5 Print your name, and sign your name as necessary.

6 Planning your time.

At the beginning of the examination see how many questions there are and divide up your time. **Roughly** plan to spend one minute for each mark.

You must **not** spend a lot of time on some questions and leave out others altogether. There are some easy marks for each question, so plan to get them.

If you have any time at the end, check over your work.

7 The examination paper may have these kinds of items:

(a) **Multiple choice items.**

For each question, suggested answers are given on your question paper.

USE A SOFT BLUNT PENCIL TO SHADE HEAVILY ONE BROWN SHAPE FOR EACH QUESTION

EXAMPLE: If you think B is the correct answer to Question 1, shade heavily the brown shape below the letter B like this:

	A	B	C	D	E
1	☐	■	☐	☐	☐

If a question has only four answers, A B C D, ignore the letter E.

Your question paper may have less than 60 questions. Use the appropriate question numbers on this answer sheet.
ANY ERRORS MUST BE THOROUGHLY RUBBED OUT USING A CLEAN ERASER.

(b) **Structured answer questions.** These are in several parts, and each part requires a few words or sentences. At the right-hand side of the paper will be the number of marks – use this to guide you as to how much to write.

(c) **Essays questions.** You **must** first plan your answer. Make an outline, for example the skin and heat loss:

Blood capillaries – widen (dilate) – heat radiation
Sweat – increase – evaporation
Hairs – flat – less air layer

Then make your notes into complete sentences.

(d) For **practical skills** questions see pages 135–40.

ALL THE BEST IN YOUR EXAMINATION!